# THE CHICKEN LADIES' GUIDE TO
# LIFE *with* CHICKENS

## Also Available from Lyons Press

*The Alpaca Handbook: The Complete Guide to Raising a Healthy Herd*

*The Backyard Gardener: Simple, Easy, and Beautiful Gardening with Vegetables, Herbs, and Flowers*

*Creating a Modern Homestead: Traditional Skills for Real, Everyday Life*

*Family Table: Farm Cooking from the Elliott Homestead*

*The Farm Girl's Guide to Preserving the Harvest: How to Can, Freeze, Dehydrate, and Ferment Your Garden's Goodness*

*The Homesteader's Guide to Cut Flowers: Growing Beautiful, Versatile Blooms for Pantry, Pleasure & Profit*

*The Homesteader's Herbal Companion: The Ultimate Guide to Growing, Preserving, and Using Herbs*

*The Homesteader's Natural Chicken Keeping Handbook: Raising a Healthy Flock from Start to Finish*

*Seasons at the Farm: Year-Round Celebrations at the Elliott Homestead*

*Welcome to the Farm: How-to Wisdom from the Elliott Homestead*

# THE CHICKEN LADIES' GUIDE TO LIFE *with* CHICKENS

## The Joys of Delicious Eggs *and* Raising a Healthy Flock

HOLLY CALLAHAN-KASMALA AND CHRISIE DICARLO,
hosts of the *Coffee with the Chicken Ladies* podcast

Foreword by TOM WATKINS

LYONS PRESS
ESSEX, CONNECTICUT

LYONS PRESS

An imprint of The Globe Pequot Publishing Group, Inc.
64 South Main Street
Essex, CT 06426
www.GlobePequot.com



British Library Cataloguing in Publication Information available

**Library of Congress Cataloging-in-Publication Data available**
ISBN 978-1-4930-9481-3 (paper : alk. paper)
ISBN 978-1-4930-9482-0 (electronic)

Printed in India

*For all of the chickens and their people.*

Chickens
Welcome
People Tolerated

# CONTENTS

# FOREWORD

Across all fields, there are not enough mentors, and there are gaps in our ability to effectively pass knowledge down. This is especially true in the poultry world. In a culture where the sum of all human history is in the palm of our hand, we have never been more blind to the simple fact that you must learn from others, figure out your path, and then go do something with what you've learned.

Holly Callahan-Kasmala and Chrisie DiCarlo are not only talking the talk but have walked the walk, and continue to do so by turning a background in animal husbandry and education into what I think is the most entertaining and informative poultry platform around today. Holly and Chrisie have a natural ability to spin a yarn and educate at the same time, firing sparks among chicken enthusiasts across the nation and beyond. Their debut book is a mixture of history and practical application—informational while being informal; comprehensive while showing a deep appreciation for the humble chicken and all things chicken-related.

You see, keeping chickens is a whole lot less about chores than it is choices. A thousand choices. Breeds, roosters, brooding, feed types, treats, meat, eggs, and, yes, daily chores. A choice to take life seasonally, to slow down, to enjoy the fruits of your labor on any scale. Chickens are a lesson in excess and scarcity on an annual basis. Those choices can seem overwhelming at first, but each step garners more confidence. Holly and Chrisie's book is a perfect step in gaining both confidence and knowledge for poultry keepers of all levels.

At its very core, *The Chicken Ladies' Guide to Life with Chickens* is a beautiful reminder of why we all keep these birds. For some, it's eggs or meat (or both). For others, birds offer companionship, help in the garden, a rhythm. My reasons for keeping chickens are numerous. Manure for the garden. Natural pest control. Kitchen scrap composters. Eggs, obviously. And if we are really honest, I appreciate the simple routine. The daily caregiving is cathartic and comfortable. Chickens fill many roles in our house, and a very important one is simple peace of mind in caring for something that takes care of you right back.

Chrisie and Holly's book lays a foundation, built on feathers, for you. And so, the next step in furthering your chicken skill set is yours: active learning. Find a mentor. Join the community. Reach out and ask questions. For all your reasons, get the chickens. And then share your knowledge with others, as Holly and Chrisie so seamlessly do on a regular basis.

Thank you, Holly, thank you, Chrisie, for putting together this engaging book and reference. For assisting in real-world knowledge transfer. For continuing to learn while taking up the mantle of mentor and spreading the many hows and whys of chicken keeping.

—Tom Watkins, president and co-owner, Murray McMurray Hatchery

*Chrisie and Holly—recording our podcast, before we had a studio.*

# INTRODUCTION

They said it couldn't be done. Everyone told us that there was no way we could talk about chickens for an hour every week. "What is there to say?" they would ask us. We answered, "Just listen." That was six years ago and we're still here, talking chickens every week on our podcast, *Coffee with the Chicken Ladies.*

This book is the direct result of our podcast. We wanted a written reference, a place where people could quickly find answers to their poultry care questions. It's also another place where we could share the celebration of chickens and the Chicken Ladies and Gents who love them. We share lots of chicken-keeping tips and info, but we also get to share the lighter side. Garden parties, collecting vintage chicken stuff, and cooking with all of those eggs take on a new dimension when we get to share stories and photos with you, our readers.

So how did we end up with a chicken podcast to begin with? That story starts in 1982. We both attended sixth grade at Deep Creek Middle School in Baltimore County, Maryland. We sat next to each other because our last names began with "Ca." Little did we know that we were destined to sit next to each other for the next forty-plus years. We've been together for everything from graduations, vacations, and weddings, to babies, new houses, and, yes, chickens.

The Chicken Lady life didn't show up right away, but unbeknownst to us, we were already preparing for it. We've always been dog and cat lovers, but after we finished college and started working, it wasn't long before we added more animals to our households.

Chrisie, a lifelong animal lover, spent fifteen years working as a vet tech / trauma nurse and client educator in a busy urban Baltimore animal hospital. Always focused on education, Chrisie trained and supervised the vet tech team, and also served as a client educator. She would carefully educate clients about nursing their animals who had just received vet care, and she also worked closely with clients who were adopting cats and dogs from the hospital.

*Chrisie's wedding, 1997.*

*Holly's wedding, 2019.*

Chrisie started her own flock of chickens thirteen years ago. Her family had moved to a new home on three acres, sparking the idea of raising chickens. She wanted to start with chicks and share the experience with her kids, teaching them the value of responsibility and the importance of food security. She left her position years ago to raise two daughters, but her focus remains on understanding chicken anatomy and health care and helping new chicken keepers. Chrisie and her girls started their flock with four Buff Orpingtons. Little did she know that she was stepping into her Chicken Lady shoes. Or boots.

Holly grew up on a small horse farm and has a lifetime of experience caring for livestock, including horses, llamas, alpacas, sheep, chickens, ducks, and geese. She worked as a history librarian for over twenty years, and even planned to write her master's thesis on livestock history (but her advisor wouldn't go for it). Holly has kept chickens for twenty-three years, focusing on heritage breeds and their history. She loves the way chickens and other poultry connect us with the past, especially with the women who have gone before us. She also loves the unique and sometimes crazy appearances that heritage-breed chickens have. She first saw Cochins on a visit to England in 1999 with her sister. When they returned home they ordered Cochin chicks, and Holly started on the road to becoming a Chicken Lady.

*Selling some of our vintage chicken goods at the Maryland Poultry Swap.*

In November of 2020, we started the podcast. We'd been watching the COVID-19 chicken boom take place that year, and we saw new chicken keepers struggling. Much of the chicken education was still based on industrial farming, and we noticed that there was not enough support for the average person who wanted to provide top-notch care for their flock. There also weren't many good resources for people whose chickens were their pets and companions. We even saw sources claiming that chickens don't need medical care, and we couldn't let that slide. After a few weeks of brainstorming, we decided to create a podcast.

Chickens are remarkable animals and they deserve the best care. They're smart, they create emotional bonds with each other, and with their humans, and they're just plain fun to have around. Not to mention the supply of fresh and delicious eggs. So, we gathered up all of the things that we love about being Chicken Ladies and wrote them into our show—and now this book. We cover breeds and history; the best ways to get started on, and to continue, your chicken-keeping journey; and plenty of information on health care and prevention. Then we layer in the good stuff: recipes and entertaining, vintage collectibles, and fancy decorations in your coops and runs.

*100,000 downloads cake, and witchy chicken tees!*

*Coffee, cake, and chicken treats, al fresco, 2025.*

We hope that this book informs, inspires, and makes you feel connected to the larger world of backyard chicken keepers. There is no one right way to be a Chicken Lady or Gent, so we encourage you to find what works best for you and your flock. This is the key to becoming successful and happy with your chickens.

We wish you many eggs, fluffy feathers, lots of chicken hugs, and a big dose of Chicken Lady joy.

CHAPTER 1

# Why Chickens?

Why chickens? Maybe the question should be "Why *not* chickens?" since there are so many great reasons to keep them. If you're reading this book, odds are you're interested in getting chickens. Or maybe you already have some, and you're looking to learn more about them. Or maybe, like us, you're just completely obsessed with chickens.

You are not alone. Chickens are more popular than ever! According to the 2025 report by the American Pet Products Association, over eleven million American households currently keep chickens. The APPA is considered the authority on pet-owning statistics, and their most recent report lists chickens as the fourth-most-popular pet in America.

Keeping chickens can be an incredibly rewarding experience. First and foremost, you have your own source of fresh and delicious eggs. Second, they're good company. As if chickens weren't wonderful enough, they also come in many different breeds, in all colors, shapes, and sizes. Some of them are better in hot weather, others in cold. Some are great at hatching and raising chicks, while others prefer to spend their time bug-hunting and foraging through your yard. Still others have super fancy feathers and look gorgeous in your garden, or they may want to spend time napping in your lap.

We're going to get into the specifics of why people choose to keep chickens, and we hope to give you a few new ideas along the way. But first we want to share a very brief history of chickens and their people.

Throughout history, women were primarily responsible for poultry care, so Chicken Ladies are a bona fide thing. Chicken Ladies have also existed across all social and economic lines. This means that chicken keeping (or "chickening," as we frequently call it) is available to everyone. We're here to help you make that happen.

*Gigi, a fancy Houdan hen who enjoys the finer things in life.*

## A Very Brief History of Chickens and Their Chicken Ladies

There is currently a lot of debate in the scholarly community about how long ago chickens were domesticated. Some scholars claim the original domestication was about ten thousand years ago; another group says eight thousand years, and most recently, more scholars claim that domestication took place around three thousand years ago. What we know for sure is that people and chickens have been living together for at least a few thousand years, and possibly much longer. This is plenty of time for chickens and their people to figure out that they belong together.

*George and Martha, a pair of Nankin bantams, an old endangered heritage breed.*

Unlike many livestock species, chickens have no wild counterparts. There is no such thing as a wild chicken. At best you can find feral chickens who have reverted to a wild state, but they tend to hang out near people, most likely trying to find someone who will feed and care for them. The chickens in Key West, Florida, are a good example of this.

Chickens as we know them are domestic descendants of wild jungle fowl. These small, colorful birds are native to South Asia and can still be found there today, in large numbers. Archaeologists in Thailand found early versions of chickens buried as "grave goods," or supplies for the afterlife, alongside other domestic livestock animals. This indicates that the birds were not wild and were kept by farmers. Studying the graves and bones and other archaeological evidence led to the conclusion that jungle fowl became domesticated because humans shared rice and other grains with them. Farmers in Asia three thousand years ago grew rice, millet, wheat, and barley, all of which chickens still eat today.

That's a simplified version of history, of course. What scientists found over and over as they studied the spread of grain farming throughout Asia and into the Middle East and Africa was the spread of early chickens alongside the farmers. Together, humans and their chickens moved out of Asia and into Europe.

Scholars cannot say with 100 percent accuracy what these early people did with their chickens. The popular assumption is that early civilizations used the eggs for food. But there is also some evidence that chickens may have been kept as pets, too. An Iron Age burial site in Austria was found to contain a surprising number of chickens. The birds were buried with their people, and archaeological isotype studies revealed that the chickens and their people had all eaten the same food. The chickens had not been slaughtered; they had been placed with care at the feet of the deceased. Even more interesting, women were buried with hens while the men were buried with roosters.

Archaeologists have uncovered other burial sites in Europe with chickens, including graves from ancient Britain. Scholars who analyzed the burial sites have concluded that chickens were treated with reverence, possibly as creatures of spiritual or religious importance, but also very possibly as pets and companions. It is believed that the Britons did not start eating their chickens until the Roman invasion. The Romans used their chickens for both eggs and meat, and were most likely a major force in spreading the consumption of meat birds throughout Europe.

*Hideo, an Australorp rooster.*

If we jump ahead to Europe in the Middle Ages, we do find some written and illustrated examples of early Chicken Ladies. The *Luttrell Psalter*, an English illuminated manuscript from the 1300s, contains an image of a woman feeding a hen and her chicks. Peasant women generally kept their own birds, while wealthier women employed dairy maids and older children to feed the birds, usually barley, and take them out for supervised free-range time. During this period, scholars agree that chickens were kept largely for eggs and only consumed as table birds for rare, festive occasions.

Chickens and other poultry continued coexisting with people for the next few thousand years. In the sixteenth and seventeenth centuries, new chicken breeds started to emerge. They were usually tied to a geographical area, like Sussex or Norfolk, and had a consistent appearance and other dependable characteristics, like a unique rose comb or good egg production, that were passed on to each generation.

*Kate Tea Biscuit, a Speckled Sussex hen, a very old British breed.*

Chickens have been present in South America for several hundred years, if not longer, but they were not present in North America until they arrived on the ships of explorers. The 1600s and 1700s saw a lot of traffic to what is now the United States. Settlers at Jamestown, Virginia, in the early 1600s had chickens with them, and more chickens and other livestock arrived on supply ships in succeeding years. To the best of our knowledge, there were no chickens on the *Mayflower* when she arrived in America, but later ships to Massachusetts did bring plenty of poultry and livestock. Dominiques and Javas were some of the earliest American chicken breeds. Dorkings and Hamburgs were brought from England, while Guelderlands, now known as Breda, were brought by Dutch settlers from the Netherlands. Under the hands of capable women, poultry keeping flourished in Colonial and Early America.

The next milestone on the chicken timeline takes place in the 1800s. Young Queen Victoria and her husband, Prince Albert, were genuine animal lovers. The queen was fascinated by birds, including domestic poultry, and soon had a large collection of fowl that she kept in an aviary at Windsor Castle. When Victoria was first gifted a pair of Cochin-China chickens, followed by a flock of fancy Brahmaputra chickens, Hen Fever took over. Everyone wanted fancy chickens like the queen's. People were paying large sums of money for pure heritage-breed chickens, the fancier the better. The Hen Fever bubble eventually popped, but by that time both men and women had become interested in keeping, breeding, and even showing purebred chickens.

In the late 1800s and early 1900s in both Europe and America, chickens kept their place on the family farm and on the show table. But the early twentieth century started to see the slow rise of larger commercial farms, including broiler farms for meat birds. The First and Second World Wars really changed chicken keeping throughout the world and sped up the creation of factory farming. The wars were so disruptive, so many lives were lost and so much farmland destroyed, that people in the Western world saw starvation and food chain breakdowns. They began searching for a way to streamline farming and to make sure that everyone had enough food. These people had the best of intentions, but the industrial farms that they developed and the hybridized laying hens that they produced had an incredibly negative effect on both chickens and the people who would normally have been keeping them. Chickens were no longer in the hands of women.

Peony, a Silver Laced Cochin hen.

Eclair, a Light Brahma hen.

*Lucy, a Leghorn hen. In the early twentieth century, white Leghorns were the industry-standard layers until the rise of hybrid layers in the 1940s.*

Mid-twentieth-century life was meant to usher in the modern era. People left their farms and small towns in search of better paying work. At the same time, suburban communities were popping up with grocery stores in close proximity. Modern families shopped at supermarkets and bought their meat and eggs there. Fewer and fewer people kept their own backyard chickens. We became disconnected from our food supply. With few people keeping and breeding them, many of the old heritage (pure) breed chickens became endangered and close to extinction. Instead of celebrating the unique and wonderful things about chickens, people began to call them stupid, smelly, and too many other pejoratives to list here. They didn't care if hens were kept in battery cages or that broilers had been bred to unsustainable standards. Chickens became a commodity instead of an animal.

In the 1970s, a major back-to-the-land movement started to revive small-scale chicken keeping. The Livestock Conservancy, back then known as the American Minor Breeds Conservancy, began the important work of identifying at-risk livestock breeds and supporting breeding efforts while educating on the importance of these domestic animals. In 1987, the Livestock Conservancy became the first organization of its kind to add poultry to its conservation list. This marked a major change in the way chickens and other fowl had been viewed through the second half of the twentieth century. Today, the Conservancy continues to educate and support breeding and work with heritage-breed chickens, ducks, geese, and turkeys. We'll talk more in chapters 4 and 6 about the pros and cons of heritage breeds.

Chicken keeping started to gain steam in the 1990s and continued to grow in the 2000s, fueled by readily available information on the Internet and the work of (very) early influencers like Martha Stewart. Many of the new chicken keepers were women, doing their best to grow food in gardens and have fresh eggs from their own hens. What we had always called hobby farming was now known as homesteading, and the movement took on a life of its own. In 2020, we saw the COVID-19 Chicken Boom. People were getting chickens at record numbers, selling out local farm supply stores and online hatcheries. And the rise of social media chronicled all of it. #crazychickenladies!

In recent years we've seen at least two more smaller jumps in small-scale chicken keeping, both corresponding to a sharp rise in egg prices. More and more people are interested in gaining some food security with chickens and a garden. In times of

*Athena, an Easter Egger hen, and Lucy, a Rhode Island Red hen. These two breeds were very popular in the early 2000s as backyard chicken keeping started to grow again.*

political unrest and economic uncertainty, having their own backyard poultry helps people feel better.

## Chicken Keeping Now

Today, there are a wide variety of reasons that people decide to keep chickens, and they are all valid. Many people start off with chickens because they want a source of fresh eggs/food. Homesteaders and gardeners may want to keep chickens as a source of manure, compost, and pest control. Chickens make wonderful pets, and there are crazy chicken people out there who keep their birds simply for companionship. Showing chickens and conservation breeding are two more reasons that people may get started with poultry.

If you're not sure exactly why you want chickens, that's okay, too. This is your opportunity to think about what you want from your flock, and what types of chicken you'd like to keep. There are no wrong answers, just opportunities for honest conversation with yourself. It's hard to have a joyful and successful journey in chicken keeping if you have convinced yourself that you need to breed lots of chickens and sell lots of eggs to make a profit, while deep down what you really want is a few fancy pet hens to look pretty and to keep your own egg cartons stocked. Or vice versa.

## Eggs

Eggs are the primary reason that people get chickens. Nothing beats the excitement of finding those first eggs in the nest boxes—it's super exciting for both kids and adults. Eggs provide a measure of food security, a sustainable way to provide food for you and your family. If you have children, getting them involved with feeding hens and egg collecting is an excellent way to help the next generation develop a responsible and compassionate connection to their food.

Chicken eggs provide a fresh source of highly digestible protein and they can be an important component of clean eating. If you're a baker, you can't get any fresher or higher-quality ingredients than your own backyard eggs. You can also create income for yourself by selling your excess eggs at a farmstand, but you'll need to check your state and local ordinances for laws, permits, and requirements.

One of the most fun things about eggs is that they come in so many different shades and colors. While they all taste the same, the different shell colors can make up a beautiful egg basket. Shades of white, cream, blue, green, pink, tan, brown, terra cotta, and dark chocolate—and some of them speckled—are all possible, depending on the breeds of chicken that you select. You can also have a variety of egg sizes, again based on your breeds.

## Gardening and Permaculture

Chickens can be a gardener's best friend. This doesn't mean that you can let your chickens loose in your garden, because they'll love eating the veggies as much as you will. But chickens produce poop, and poop eventually breaks down into a major component of healthy soil. You can also create compost with soiled coop bedding and chicken-run substrate. Some gardeners and permaculture practitioners are so good at this that they rarely if ever need to buy or bring in outside compost or soil.

Chickens can also help in the garden by turning over spent beds at the end of the season and clearing beds in early spring. Because they're such good foragers, chickens can also help you turn over your compost pile, allowing the microbes to do their job faster and create usable compost more quickly. For a fun treat, share spent or bolted veggies with your flock when you're ready to replant a bed. They'll happily gobble down the flowers, leaves, and small stalks.

◀ *Spicy, a Rhode Island Red, enjoying some garden time.*

## Pets and Companions

Contrary to what you may have heard, chickens make excellent pets. There's nothing we like better than settling down in a comfy chair with a cup of coffee and a warm hen napping in our lap. Chickens have the ability to form emotional bonds with their caretakers, as well as with other animals. Some breeds are more gentle and friendly than others, but the more time you put in with them when they're young, the easier it is to win their love. Some breeds are also more appropriate for children than others; we'll talk more about this in chapter 6, when we discuss breeds. Regardless of what kind of chickens you decide to get, children can benefit from learning safe handling and how to meet the physical and mental/emotional needs of chickens so that the birds can thrive along with your kids.

Our podcast listeners tell us that one of their favorite things to do after a long day at work is head out to the chickens and watch them dig and scratch. This Chicken TV is incredibly relaxing and has been proven to bring down blood pressure and increase feelings of well-being and contentment. Once you start keeping chickens, you'll quickly realize that they are all individuals with unique personalities, and this will make the whole experience even more fun and interesting.

Keeping chickens as pets also allows you to keep your birds in some nontraditional ways. Some folks choose to keep only roosters as pets, because they're so sweet and affectionate if they don't have hens to look after. Other people choose to keep house chickens, often special-needs birds that cannot survive in a regular flock. The most

*Gertie the Welsummer, a special-needs hen.*

important thing to remember is that any chicken can be your pet. You are free to make pets out of your laying hens, roosters—any or all of your birds. No one gets to tell you that a chicken is only for supplying food. You get to make that decision yourself. As Chicken Ladies, we make pets out of all of them. They're our beloved birds, and they remain with us for their entire lives.

There's one more benefit to keeping pet chickens that not everyone knows about. It is the Chicken Lady Network. This is an informal group of people who *love* their poultry. These Ladies and Gents are friendly and generous with their time and assistance. If you are becoming a Chicken Lady, you want to tap into this social network. You may find some brand-new best friends this way. And nothing is better than having a friend or two to talk over all of the details of chicken keeping.

## Show Chickens

While we don't show any of our poultry, showing chickens is a popular pastime. From 4-H kids showing chickens and eggs at the county fair (which we highly recommend!) to American Poultry Association members, folks enjoy showing off their birds. The American Poultry Association is an organization that serves as the major governing body for US poultry shows, as well as the stewards of the breed standards for domestic poultry. There is a lot to see and learn at chicken shows, so it may be worth checking out a local show if you think you might be interested in poultry exhibition. You will see some of the fanciest chickens ever!

## Conservation Breeding

The final reason that people keep chickens is a very important one, and not something that everyone can do. Breeding chickens for conservation takes special dedication and the right infrastructure to support a breeding operation. Heritage-breed chickens, with their long histories, fancy feathers, and other unique characteristics, often need devoted breeders to keep their lines going and safeguard important genetics. Some people start off with a few pet chickens and feel called to begin breeding. It is not the easiest path, but we are all indebted to these special chicken keepers. They protect and preserve precious biodiversity that could someday be the means of saving our food supply.

*Scenes from the 150th Ohio National Poultry Show in November 2023.*

Bantams are tiny chickens that are roughly one-third to one-fourth the size of standard chickens. Most bantams weigh between 1 and 2.5 pounds. The oldest bantam breeds are known as "true bantams"—and only exist as bantams—while other bantam breeds are tiny replicas of large fowl.

*A freshly hatched Nankin bantam chick takes a nap after pushing his way out of the shell.*

## Frequently Asked Questions

When you're just starting out with chickens, you'll naturally have a lot of questions, because keeping chickens isn't a life skill that has been passed on to most of us. We remember what it was like to start out and not be able to find answers to our most basic questions. So, we've compiled this list of FAQs that we've received from our podcast listeners and followers over the years. We hope you find these helpful!

- **Do you need a rooster for your hens to lay eggs?** No, hens will produce eggs start to finish without a rooster around. However, eggs cannot be fertilized and cannot produce chicks if there is no rooster in with the hens.
- **Why do you give your hens oyster shells?** Laying hens need an abundant form of long-acting calcium to maintain their health—even more than what is found in

food formulated for laying hens. Oyster shells act as a slow-release form of calcium in their bodies, helping them form strong eggshells and maintain bone density.

- **How old are hens when they start laying?** Depending on their breed, most hens will begin laying eggs at five to six months old. Some heritage-breed hens will begin to lay eggs when they are older, often between six and eight months.
- **How long do chickens live?** Heritage-breed chickens can live anywhere from eight to twelve years, sometimes longer. Hybrid laying hens tend to have a shorter life span, anywhere from two to six years.
- **Are chickens smart?** Scientists have determined that chickens are quite intelligent. Their brains are small but powerful, like a computer chip. They have cognitive reasoning skills, excellent memories, and the ability to problem-solve. Chickens are amazing!
- **What do I feed my chickens? Can I just feed them with table scraps? What about scratch grains?** Chickens thrive on a balanced diet that gives them their recommended daily allowance of the vitamins and minerals they need. A diet consisting of only table scraps may not contain the nutrients they require, and could lead to soft-shelled eggs, vitamin deficiencies, and other health problems. Scratch grains (supplemental whole grains like corn, wheat, barley, and oats, fed in the evenings or during cold weather) can be fortified, but rarely contain everything that your chickens need every day. We recommend feeding your flock a balanced daily ration of high-quality chicken feed. In the afternoon and evening, you can supplement with scratch grains or table scraps.
- **Will I save money on eggs by keeping chickens?** If you are only planning to keep chickens for a short amount of time, you probably won't save any money. The longer you have chickens, and the more years you have healthy hens producing eggs for you, the more your initial investment pays off.
- **How many years will my chickens lay eggs?** The healthier and better cared for your hens are, the longer they generally lay. You won't get tons of eggs from your senior girls, but they tend to give a handful each spring. We both have nine-year-old hens who are still laying. Artificially lighting your coop and forcing your hens to lay in winter will shorten their laying years.

- **Why is my chicken losing feathers?** There can be several reasons for feather loss, but the primary cause is the chicken's annual molt. Every year in late summer and into fall, chickens will drop the majority of their feathers and grow new, healthy feathers to keep them warm in the cold months.
- **Can chickens get sick?** Yes, as with any other living creature, chickens can get sick. Like many flock/prey animals, chickens hide their symptoms well, so it's up to us as their caretakers to keep an eye on them and learn their behaviors so we can spot problems early.
- **Can I give my chickens apple cider vinegar every day?** No, every day is not a good idea. Apple cider vinegar is best used no more than once per week. The easy-to-remember formula is 1 tablespoon in 1 gallon of water once a week. Using ACV too frequently can cause imbalances and create health problems, including diarrhea and damage to the lining of the digestive system. It works well as an acidifier for the chicken's crop and digestive system, lowering the pH and creating an inhospitable environment for unwanted microbes. ACV is best used as a preventive measure.
- **Can you eat a fertilized egg?** Yes, you can eat a fertilized egg. You do not want to eat an egg once it has started developing. If you keep a rooster, it's a good idea to collect eggs daily so your hens don't have a chance to go broody and start sitting on a clutch of eggs.
- **What is a broody hen?** A broody hen is a chicken whose hormones have activated her desire to hatch a clutch of chicks. She will stop laying eggs and sit on the nest for the twenty-one days that it takes for chicks to develop and hatch. Once hatched, she will care for the chicks until they are old enough to be on their own.
- **Do I need a coop and run for my chickens?** You will need a coop and run for your chickens if you want to protect them from weather and predators, and if you want them to have a secure place to lay their eggs. We do not advise keeping chickens without predator-safe coops and runs that also keep them dry and safe from the elements. Some people choose not to have a run and to free-range their chickens every day, but you will lose some (or all) of your chickens to predators this way.
- **What chickens lay the most eggs?** In general, hybrid laying hens will produce the most eggs, but they also tend to have shorter life spans and more reproductive issues. Among the heritage-breed chickens, the top layers are Leghorns, Rhode Island Reds, and Australorps.

## Pros and Cons of Keeping Chickens

Before you run out and buy your first chicks, we encourage you to take a moment to read through the pros and cons of chicken keeping. While we think anyone who wants chickens should be able to have them, we also want future Chicken Ladies and Gents to make informed decisions and know what they're getting themselves into. Chicken keeping doesn't have to be hard, but it will impact your lifestyle in certain ways and it's best to be prepared.

| **Pros** | **Cons** |
| --- | --- |
| Fresh eggs | Cost of coop/run, feed, and vet/medical care |
| Producing your own food | Eggs are seasonal, not available all year. |
| Free fertilizer | Cleaning coops takes time and extra work. |
| Garden help | Unsupervised chickens can destroy a garden. |
| Feathery friends, unlimited hugs | It hurts when you lose them. |
| Connecting kids/adults to their food source | Not every kid wants to keep chickens. |
| Clean eating | Sometimes you have too many eggs! |
| Chicken Lady life | Chores 365 days, a.m. and p.m., in all weather! |
| Seasonal living | Sick or injured chicken = Stress |
| De-stressing | |

## Keeping Chickens vs. Ducks vs. Geese

Chickens aren't the only source for wonderful, fresh eggs. Ducks and geese are also popular with backyard poultry keepers. All three types of birds have their own requirements. Chickens and ducks lay more eggs than geese, but geese make terrific pets and alarm systems if you don't mind the noise. None of them are better than the others, just different. Chrisie keeps chickens, while Holly keeps chickens, ducks, and geese.

| | Chickens | Ducks | Geese |
|---|---|---|---|
| **Egg Size** | Small-XL | XL-Jumbo | Huge (=3 hen eggs) |
| **Annual Egg Counts** | 100-350 | 100-350 | 15-80 (seasonal) |
| **Number of Breeds** | 150+ | 40+ | 12+ |
| **Water Needs** | Drinking water | Drinking water with food; water for dunking head; swimming water necessary | Drinking water with food; water for dunking head; swimming water appreciated |
| **Life Span** | 8-12 years | 8-12 years | 20+ years |
| **Living Arrangements** | Coop with roosts and nest boxes | Sleep and nest on the floor | Sleep and nest on the floor |
| **Personality** | Gentle-Spicy | Gentle-Mildly Spicy | Gentle-Spicy |
| **Foot and Mouth** | Beak and claws | Bill and webbed feet | Bill and webbed feet |
| **Body Sizes** | Bantam-Large Fowl | Bantam-Large Fowl | Large, no Bantam |
| **Cold Hardiness** | Some breeds, very | Very | Very |
| **Heat Hardiness** | Some breeds, very | Okay with water | Okay with water |
| **Easy to Acquire?** | Easiest | Easier | Seek out breeders and hatcheries |

CHAPTER 2

# All About Eggs

The incredible, edible egg! Does anyone remember that jingle from the 1970s and '80s? It was the slogan for the American Egg Board. Eggs are an important part of our lives, probably more important than most of us think. Not only do we eat them on their own several times a week, they're also used in baked goods and lots of other food preparations. Planning a birthday party? There are eggs in the cake. Afternoon tea? Egg salad sandwiches and meringues. Graduation party? Eggs in the pasta salad. Fourth of July? Deviled eggs. Easter brunch? Eggs everywhere!

For something that is so important to us, most of us don't know much about eggs. When we started our *Coffee with the Chicken Ladies* podcast, we set out to learn, and share, as much about eggs as we could. After all, they were the primary reason why we both got chickens.

Let's take a closer look at all of the amazing things about those eggs.

## History

Humans have been eating eggs for millions of years. Early humans would have eaten eggs that they foraged from wild birds, but as they settled on land and started farming, domesticated poultry would have become their egg supply. In addition to chickens, people consumed eggs from geese, ducks, turkeys, quail, pigeons, Guinea hens, and even peacocks and ostriches.

Food historians believe that people began baking with eggs—most likely in bread, cake, and pastry recipes—in ancient Egypt and Rome. It is also very likely that eggs were being used to cook in ancient China around the same time. Simple egg fried rice is believed to be one of the oldest Chinese egg dishes; evidence suggests it could date back to the third century BC.

In 2016, archaeologists on a dig in Aylesbury, Buckinghamshire, in the United Kingdom, found a 1,700-year-old intact chicken egg in an area once inhabited by Romans. The egg is believed to have come from chickens that the Romans had brought to Britain. The archaeologists noted that there were other formerly intact eggs in the dig site, but they were broken during the excavation.

Custard, a mixture of eggs and milk, is a foundation recipe for so much of our modern cookery. It is believed to have been invented in Rome over two thousand years ago. The recipe first appeared in *Apicius*, a book of recipes written by Roman gourmet Marcus Gavius Apicius. He lived in the first century AD, and his custard recipe calls for milk, eggs, and honey.

If we fast-forward to Medieval Europe, we start to see eggs used more regularly in recipes. As we mentioned in chapter 1, we know that women in this period were the primary caretakers of poultry, so it stands to reason that they would incorporate their abundance of eggs into their cookery. By this point in time it is safe to say that chicken eggs were regularly consumed in most of the world, but they were not yet on the menu in North America. Chickens were present in South America, specifically Chile, but it isn't clear when the Indigenous people began cooking with their eggs.

In 1493, Christopher Columbus brought chickens to the Caribbean islands on his second voyage to the Americas. These chickens remained on the islands and were an important source of food for explorers and sailors over the next two hundred to three hundred years.

In 1606, the Virginia Company in London, England, outfitted three ships to cross the Atlantic and settle in the new colony of Jamestown, Virginia. These and later supply ships carried chickens to the American colonies. Chickens and other poultry were

*Dolley Madison, a Dominique hen. Dominiques were the first breed of chicken developed in America.*

often not considered important enough to be listed along with larger livestock, so they are not always mentioned on ships' bills of lading. However, by the 1700s there were thousands of chickens spread across the territory that would later become America.

If you look at the recipes that were common in Colonial America, you'll see a lot of eggs used in various ways. Corn fritters would use eggs as a binder, while pancakes used eggs for both binding and lift. Sauces were thickened with the yolks of hard-boiled eggs, and cakes of all sorts used eggs in lots of different ways. Around this same time, eggs started becoming more popular as a breakfast food. Buttered eggs was a dish popular in Colonial/Early America—essentially, scrambled eggs with a lot of butter.

America largely kept to the tradition of women being the primary caretakers of chickens and other poultry, and this gave rise to a unique and widely accepted

The Incredible
Egg Scale tm
30
27
XL
24
L
21
M
S
18
2
57
$1\frac{3}{4}$
50
$1\frac{1}{2}$
43
OZ.
DOZ
OZ.
Egg
gram
Egg

*Egg scale.*

economic opportunity: egg money. Women were allowed to keep any money they made selling eggs from their own birds. Enslaved people were generally allowed to keep their own chickens and sell eggs and meat as well. While eggs weren't selling at high prices—according to *Martha Washington's Booke of Cookery*, a dozen eggs cost 3 pence—this practice did put money into the hands of members of Early American society who had no other economic opportunities. The tradition of egg money persisted up through the early twentieth century, until it essentially died out after the World Wars and the shift to industrial egg farming.

## What Are Eggs Made Of?

Eggshells are composed almost entirely of calcium carbonate crystals. This is the same material that makes up seashells. These crystals are basically nanoparticles that connect and form a hard calcite shell. In order for a chick to develop, the egg needs to allow for the exchange of gases, so oxygen and moisture are present inside. This is possible because each eggshell contains thousands of tiny pores (generally between 7,000 and 9,000).

The egg white is 90 percent water and contains almost no fat. The other 10 percent is bundles of protein. The egg white is surrounded by a membrane that protects the yolk—and the developing embryo, in fertilized eggs—from bacteria and helps hold in moisture. This membrane layer is made of fibrous proteins such as keratin.

Egg yolks are composed of various fatty acids, cholesterol, fat-soluble vitamins (A, D, E, and K), and xanthophylls (the compounds that give yolks their color and belong to the same family as beta-carotene). This is why feeding your chickens foods that are high in beta-carotenes can help make egg yolks more orange. Marigold petals are a popular chicken-food addition that can make egg yolks a deeper orange color.

The other thing present in an egg is the air cell, or air sac. This is necessary for an embryo and developing chick's survival. The air sac remains when you hard boil an egg—it's that empty space under the shell that makes it easy to start peeling.

## How Are Eggs Formed?

Understanding your hen's reproductive system is crucial to identifying any egg-related problems she may have. Her whole system isn't long or complicated, but it really fulfills a lot of functions along the way. It's also completely amazing!

Hens (and most females of any species) are born with all of the eggs they'll ever lay. A hen has only one ovary, located about halfway down her back, roughly between the neck and tail. The reproductive system is basically a twisted tube that is usually 25–27 inches long. The ovary contains her ova, which are essentially the developing yolks. When a yolk is ready, her ovary releases it into the oviduct. This generally happens every one to three or four days, depending on age, breed, and time of year.

Once the developed yolk is released into the oviduct, the rest of the magic starts to happen. The oviduct is lined with a few different kinds of glands that each have different jobs. The first stop is the infundibulum. This is where the egg develops the chalazae—the white, stringy fibers that hold the yolk in place—and where the egg would be fertilized if the hen had mated. The egg stays at this stop for about fifteen minutes or so before it moves on to the magnum.

The magnum (derived from the Latin for "large") is the largest section of the oviduct, and the egg remains here for about three hours. This is where the albumin, or egg white, is secreted and forms around the eggs. From the magnum, the egg moves into the isthmus. The isthmus, which is about 4 inches long, keeps the egg for about an hour and fifteen minutes while it forms both the inner membrane and the soft outer shell membrane.

The next-to-last stop for the egg is the shell gland. This gland is about 4–5 inches long and forms the hard eggshell. This takes a while, and the egg can remain in the shell gland for twenty-plus hours while the hen secretes the calcium carbonate necessary for a nice, hard shell. This highlights the importance of plenty of calcium in your hen's diet! She can use as much as 10 percent of the calcium in her bones for eggshells if necessary, so it's better for her health to keep her supplemented. The shell gland is also the spot where pigment is deposited on the shell.

The final stop on the egg's journey is the vagina. The egg is fully formed by this point, but this is where the bloom—a bio-film that serves as a natural shield against contaminants—is deposited on the shell to protect the pores. If the hen mated, the sperm is stored here until it migrates up into the infundibulum. The egg also turns here. It travels through the oviduct pointy end first, but turns wide end first before the hen lays the egg.

The egg's journey through the oviduct takes about twenty-five to twenty-six hours total. A new egg is released and starts the cycle over about thirty minutes to an hour and fifteen minutes after the previous egg has been laid!

## What Determines Shell Colors?

If you've spent any time on social media over the last several years looking at chicken keepers' content, you will have encountered #rainboweggs. "Rainbow eggs" as a concept just means keeping chickens that give the biggest variety of eggshell colors possible. This trend has its roots in the work of Martha Stewart. In the 1990s and early 2000s, Martha featured the beautiful chocolate-brown eggs and turquoise-blue eggs from her French Copper Marans and Ameraucana chickens, among other breeds. We remember seeing these eggs in the *Martha Stewart* magazine and on her network television show around twenty years ago. Martha popularized colored eggshells at a time when few people were thinking about them.

Fast-forward to the 2020s, and everyone is all about the colored eggs. You will see a huge variety of colors on social media, some natural and some with the help of photo-editing filters. In reality, there are only four egg colors: white, brown, blue, and green.

*Eggs from Chrisie's Cuckoo Maran hens.*

All eggshells would be white without the addition of pigment. The pigment secreted by brown egg layers is called protoporphyrin and is derived from heme, the compound that gives red blood cells their color. This pigment is produced in the cells lining the hen's oviduct. The results range from a light, creamy-tinted white to the darkest chocolate brown.

Blue eggs are caused by biliverdin, a precursor to bilirubin. The blue pigment is also applied to eggs in the hen's oviduct, but unlike the brown pigment, the blue permeates the shell completely. Only chickens with genes linked to South American Araucanas produce the blue shell pigment. Green eggs are simply blue eggs that have had the brown protoporphyrin pigment applied to the outer shell.

Every other color is a matter of shade (especially brown) or is a result of the bloom, which can add a further layer of color to an egg, making a brown egg look dusty pink or plum-colored or a green look darker or lighter. Washing or wetting an egg will remove the bloom, not only changing the color but also removing the egg's natural protection.

## Nutritional Content

Eggs are incredibly nutrient-dense, easy to digest, and an excellent source of high-quality protein. In other words, they are an important food source for humans.

We're going to break down the nutrients in an average large-sized egg. There has been independent research conducted that proves eggs from backyard hens are even more nutrient-rich than store-bought eggs, but for our baseline, we're using statistics from the American Egg Board.

The average large egg purchased from a grocery store has 70 calories, 5 grams of fat in the form of healthy polyunsaturated and monounsaturated fats, 0 carbohydrates, 0 fiber, and 6 grams of protein.

Eggs do contain about 70 mg of sodium and 185 mg of cholesterol. According to the Harvard School of Public Health, research shows that for most people, cholesterol in specific foods has a smaller effect on blood levels of total cholesterol and harmful LDL cholesterol than does the overall mix of fats in our diet. In other words, eggs and the fat they contain are generally pretty safe to eat and not necessarily a contributor to levels of "bad" LDL cholesterol.

Eggs also contain a plethora of vitamins and minerals. They are a good source of vitamin D, vitamin A, the B vitamins, niacin, iodine, phosphorus, zinc, selenium, and choline. They are also a moderate source of calcium, iron, and vitamin E. Plus, eggs contain the antioxidants lutein and zeaxanthin. Antioxidants protect our bodies from cellular damage by fighting dangerous free radical cells. That's a lot of healthy stuff packed into one little egg! And this is one of the many reasons why we hug our chickens every day.

## How Many Eggs Do You Need?

Calculating the number of eggs that your household uses per week can be as simple as keeping track of how many dozens you currently use. Statistics say that most four-person households use about two dozen eggs per week, though we feel like the number is closer to three dozen per week for a family that both eats and bakes with a lot of eggs. In either case, once you know how many eggs you go through a week, the next thing you'll need to figure out is how many hens to get.

There's a big variable here, and that is the fact that different breeds of chickens lay different numbers of eggs. Some breeds are very good layers, but that is not the only reason to choose a chicken breed. You need to take many other factors into consideration to find the chickens that best meet your needs and fit your lifestyle. In chapter 6, we give you the laying stats of our favorite heritage breeds, along with lots of other pertinent information. For now, just figure that if you want six to nine eggs per person per week, you'll need anywhere from four to eight hens, depending on breed.

The other big variable here is the fact that most hens lay fewer eggs as they age. Many breeds start to slow down at around two years, though there are plenty of exceptions to this rule. For the Chicken Lady, this is an excellent opportunity to add some more birds to her flock every few years. But not everyone wants to increase their flock, so be mindful of longevity when you choose your breeds.

## Collecting Eggs

Collecting eggs is one of the highlights of keeping chickens. While the very first eggs your girls lay are truly exciting, we're happy to report that collecting those eggs never, ever gets old. It's a source of pure joy every single day to pop open the nest-box doors and find lots of little gifts waiting there. But, in order to keep things running smoothly, there are some best practices when it comes to managing your daily egg supply.

We strongly recommend that you collect eggs daily, even twice a day if you have the opportunity. There are several reasons for this. First, it helps prevent any damage to the eggs as they sit in the box throughout the day. They can be damaged by another hen laying her egg directly on top, and sometimes the chickens will peck the eggs and crack or put holes in them. Predators such as snakes, rats, and raccoons will happily steal and eat chicken eggs. On rainy days, collecting eggs promptly will cut down on dirty hen feet getting mud on clean shells. And in sub-freezing weather, eggshells will develop cracks if they're left in the nest boxes long enough to freeze.

If you have kids, collecting eggs is one of the best ways to get them involved in chicken keeping. Even preschoolers can help collect eggs if you manage the nest-box lids and the basket for them. You may end up with a broken egg or two, but the trade-off is helping kids develop fine-motor skills and hand–eye coordination. Egg collection also helps give kids a connection to their food supply, teaches them the daily responsibilities of animal care, and gives them the satisfaction of having contributed to gathering food for the family. If you're worried about your little ones in with the chickens, you can find nest boxes that open from outside the run, keeping everyone in their own safe spaces.

We mentioned chickens pecking and cracking eggs. Some birds become egg eaters. There are a bunch of reasons why this can happen, including boredom, hunger, nutritional deficiencies, nest-box dramas, and simple curiosity. One of the easiest ways to cut down on egg eating is to collect eggs as often as possible. You can also purchase nest boxes that slope downwards so the eggs roll away from the hens after they're laid, keeping them clean and safe from marauding egg eaters. You may hear that you cannot cure an egg eater, but we have done it. It just takes consistency and some patience. We also disagree with the notion that egg eaters should be culled from your flock. They can be cured with some work, but if you don't mind the occasional egg being consumed, then there's no reason to remove them.

Strong, healthy eggshells make it harder for egg eaters to indulge, while soft shells provide easy access. Make sure your hens have free-choice crushed oyster shell available at all times. A soft layer of bedding or a nest-box liner also protects eggs from cracking when they drop.

There are a lot of fun vintage options for egg collecting vessels. You can use anything from vintage baskets to fancy aprons. While we love to collect these things, we generally use the most convenient carriers that happen to be around. Chrisie's daughters carry in her eggs, while most days Holly just uses her T-shirt.

## Cleaning and Testing Eggs for Freshness

It may surprise you to learn that it is safe to store your fresh eggs on the kitchen counter for a few weeks rather than in the refrigerator. Remember earlier when we talked about the bloom that is applied to each egg before the hen lays it? The bloom is a biofilm that coats the pores in the egg to stop bacteria from contaminating it. This very cool feature most likely evolved as a way to protect chicks growing in the shell. It takes twenty-one days for a chick to hatch, roughly the same amount of time the bloom keeps an egg fresh.

*Holly holds a small egg carton in a London shop. Eggs in the United Kingdom are stored on the shelf.*

In the United States and Canada, eggs sold to the public are required to be washed, which removes the protective bloom, and then refrigerated right away to prevent contamination. But the rest of the world sells eggs without washing and refrigeration, relying on the bloom to do its work. If you decide to sell your eggs in the United States, you'll not only have state and local ordinances to follow, but you'll also have to wash, grade the eggs for size and quality, and then refrigerate them.

Of course, if it makes you more comfortable, you can wash your eggs and refrigerate them immediately. While we both store our eggs on the counter, we will wash very dirty eggs or eggs that have had poop happen and pop those into the fridge. It comes down to your individual preference. We do wash the eggs that we store on the counter right before we use them. It's the best way to remove any lingering bacteria from the eggshell. Our absolute favorite way to wash eggs is with a high-quality silicone scrubber and an enzyme wash.

Are you wondering how to figure out if an egg is old or spoiled? The easiest way to do this is by floating your eggs. You'll want a bowl or mug that's deep enough for an egg to float. Fill it with cool water and carefully drop the egg in question into the vessel. If the egg floats, it's probably contaminated with bacteria and should be discarded. If it's a little old but still good, the egg will stand on its end but remain underwater. A very fresh egg will fall to the bottom and lie on its side. And it's always a good idea to smell each egg that you break open. It's rare, but every so often you'll find a bad egg that still falls to the bottom and passes the float test. Better safe than sorry!

***Pro tip:*** Crack each egg into a small ramekin or bowl before adding to your recipe. This way, if you do get a bad egg, you won't have spoiled the whole dish.

*The fresh egg sinks while the bad egg floats.*

## Storing and Displaying Your Eggs

There are lots of ways to store and display your eggs—this is an area where Chicken People can get creative. Some of them are practical, others fancy and fun. The only objectives here are to keep your eggs securely contained and maybe show them off, if that's what you like. We do like showing off our eggs. A basket of fresh eggs in the kitchen is the best decoration!

The most practical storage and easiest to get your hands on is the regular one-dozen egg carton. These can be paper, plastic, Styrofoam, or even reusable rigid plastic. Lots of people save these and are happy to give them to you once they know you keep chickens. We use these cartons every day for eggs that need to be stored in the fridge or for eggs that we're gifting to someone.

Egg baskets are another very popular way to store eggs. They're double-duty, because you can collect with them but they also look very nice on a table or counter with your eggs placed inside. Some of the most popular are made of wire, either in a traditional basket shape or in the shape of a chicken.

Over the past several years, houseware companies have begun offering ceramic dishes made specifically for holding eggs. They're strong and sturdy, perfect for storage on your countertop. They come in all colors and designs and add a fun element to your kitchen.

*Hen-on-nest dishes were developed in the late 1800s. Large dishes are perfect for egg storage.*

We like to have fun and switch it up. We'll use deviled-egg plates, fancy-footed bowls, vintage crystal that we found at the thrift shop, the inside of large "hen-on-nest" dishes, wooden crates, and other unique or artisan-produced containers. As long as it holds your eggs safely, anything goes. And we love using vintage egg cups to show off special eggs, or for egg-centric tablescapes.

## The Many Culinary Uses for Eggs

As serious bakers, having our own fresh egg supply is very important to us. Eggs serve several different functions in the kitchen, including flavoring, protein, binding agent, leavening and lift, glaze, and the delicious pillowy glory that is freshly whipped meringue.

Most of us eat and enjoy eggs regularly. Scrambled, fried, poached, or hard-boiled eggs are the usual forms. But the taste of eggs also shines in custard-based dishes and desserts. Delicious bread pudding is a great example of a sweet custard, while a frittata is a savory egg custard. And French toast is one of our absolute favorite uses of custard!

As we mentioned earlier, eggs are an excellent source of highly digestible protein. Anywhere they're used in a dish, you're getting clean protein and lots of essential vitamins and minerals. There is a Sicilian tradition of adding protein to pasta by incorporating hard-boiled eggs; sometimes they're in the sauce, and sometimes they're served on the top. It's delicious! Vegetable stir fry and rice also gain protein from the addition of scrambled egg.

Eggs are excellent binding agents. You can use them to hold together anything from veggie burgers to cookies to birthday cakes. The sticky egg whites bring all of the ingredients together and bind them as the proteins expand from cooking. In baked goods, the eggs help water and oil and the rest of the ingredients mix together into a cohesive batter that bakes evenly. Egg whites are also excellent as a kind of edible glue. They're one of the primary ingredients in royal icing (the hard white kind), and they can also be used as a base for sugared flowers and fruit.

While we're on the subject of cakes, eggs are commonly used as leavening agents to help batter rise as it bakes. So many baked goods—including cakes and cupcakes, soufflés, muffins, brownies, cornbread, and even some biscuits—owe some of their rise to eggs. Beating air into the eggs creates volume in the batter and contributes to a light and fluffy texture.

Another culinary use is glaze or egg wash. A mixture of beaten egg or egg whites and water, milk, or cream can be brushed onto pastry before placing it in the oven. This gives baked goods a beautiful brown and shiny coating.

*Holly's fresh apple pie looks picture-perfect thanks to a generous egg wash.*

## Preserving Eggs

We've gotten used to having eggs available year-round. So, why do we need to preserve them? It's because eggs are actually a seasonal food. With the exception of hybrid laying hens and first-year pullets, chickens do not lay eggs year-round. They naturally stop in the late summer through late autumn, to give their reproductive system a break as they molt and grow new feathers.

The amount of time that your hen is "off of lay" will vary, but it averages around five months. Depending on your breeds, some hens start laying again in late December and early January; these hens are usually referred to as winter layers. As we move into spring, the rest of your flock, including those Leghorns and warm weather–loving Mediterraneans, will join in. Before you know it, your flock is back into the swing of things and you have tons of eggs once again.

But how do you get eggs during those off months? We prefer not to buy them from the grocery store for a variety of reasons. Sometimes you're lucky enough to live near someone who has first-year pullets laying through the winter. (A chick born in spring will generally start laying in fall and will produce eggs through that first winter. In her second year and every year after, she will stop laying in order to molt.) But if there's no good egg source near you and you'd prefer eggs from your own hens anyway, then you will want to preserve some of your supply during the heavy production months.

Egg preservation methods fall into three groups. You can freeze them, you can dry them, or you can use something to block the pores on the egg to stop contamination. Each method has its pros and cons.

### Freezing

Our preferred method is freezing the eggs. Unfortunately, you cannot freeze them whole, and the yolks don't freeze well on their own, though whites freeze just fine. We want our eggs for holiday baking, so the method we prefer is scrambling the yolks and whites together and then freezing them. We have found the scramble/freeze method gives us the best results.

You can freeze your scrambled eggs in any size measurement that you want and they work in any recipe where the eggs will be mixed or beaten. Frozen eggs are best

*Freeze one beaten egg in each muffin cavity.*

used within six months; discard them after one year. When you want to use your frozen scramble, just defrost in the fridge overnight or under cold running water. Once they're defrosted, you need to use them right away. Don't try to re-freeze them, as the risk for bacterial contamination has been introduced.

| Pros | Cons |
|---|---|
| Easy to do | Eggs cannot be separated, so they cannot be used for meringues or any recipe that needs egg whites whipped. |
| Doesn't require extra equipment | If you lose electricity and don't have a backup for your freezer, you may lose what you've preserved. |
| Doesn't take up a lot of space | |
| Consistent results | |
| Long-term storage, up to one year | |
| Very safe | |

## Drying

The second type of preservation is drying your eggs. There are a few different ways to do this, including dehydrating, freeze-drying, and salt-curing egg yolks. You can also dehydrate or freeze-dry egg whites on their own.

To dehydrate, you'll need to scramble the eggs and pour them onto your dehydrator trays in thin layers. Once they're completely dehydrated, you can process them into a powder. You will want to store them in a sealed, oxygen-free container in a cool environment. Unopened, they're good for five years. Once you open the container, you'll need to consume the eggs within a couple of weeks. Reconstitute the eggs by mixing 1 tablespoon of egg powder with 2 tablespoons of water or milk.

| Pros | Cons |
|---|---|
| Very long-term storage | Requires a dehydrator |
| Can fit a lot of eggs in a small space | Taste and texture are not the best. |
| Can be eaten as scrambled eggs | Reconstituted eggs do not work well in baked goods. |
| Egg whites can be used to make meringue or royal icing. | Loss of some nutrients |

The best method for freeze-drying eggs is to scramble the eggs and then freeze them before you move them to the freeze-dryer. The freeze-dried eggs can then be crumbled or processed in a blender or food processor. The resulting fine powder can be stored in an airtight container and will last several years in storage. Like dehydrated eggs, freeze-dried eggs can be reconstituted by mixing 1 tablespoon of powdered egg and 2 tablespoons of water or milk.

**Pros**

Retains more nutrients than dehydrated eggs

Improved texture and taste

Can be stored for years in an oxygen-free container

Work reasonably well in baking recipes

**Cons**

Freeze-dryer equipment is expensive

Requires extra steps

People have also been salt-curing egg yolks for more than a thousand years, relying on salt's ability to remove moisture through osmosis. To try this drying method, you need a dish or container that is deep enough for about 2 inches of salt. Pour an inch into the dish, carefully place your yolks on the salt without letting them touch each other, then cover with another inch of salt. If any of the yolks break, you will need to remove them. Place the dish in the refrigerator for a week. After a full week you can remove your dish and uncover the yolks. Brush off the excess salt and, using a strip of cheesecloth, wrap and tie off each yolk individually so they're not touching within the bundle. Return to the refrigerator for another week. After the second week of curing, the yolks keep for about a month.

◀ *Salt-cured egg yolks taste delicious grated over soup.*

| Pros | Cons |
| --- | --- |
| Requires very little equipment | Limited uses for the yolks, and leaves the whites to be used elsewhere |
| Inexpensive and easy to do | Don't keep long |
| Good taste | |

## Sealing the Pores

The final type of egg preservation relies on sealing the pores on the eggs. There are a variety of ways to do this, including submerging the eggs into a crock or bucket of pickling lime or sodium silicate—a process known as water glassing—or blocking the pores with mineral oil, lard or shortening, or even butter. This method requires the freshest, cleanest eggs and is not effective if the bloom has been washed off. Otherwise, you risk sealing bacterial contaminants into the shell. Some scientists warn that because you can't see bacteria, this method has a higher risk level than any of the others. Water glassing has been in use since the early 1800s, but is not currently recommended by the FDA.

Store the sealed eggs in a cool, dark place. If you're using this method, the safest option is to use the eggs within six months to a year.

| Pros | Cons |
| --- | --- |
| Inexpensive and relatively easy | Possible bacterial contamination |
| Moderately long storage life | Crocks and buckets require more storage space. |
| Can be used in all of the regular ways | |

mrs.wages
PICKLING LIME
For Extra Crispness
70 YEARS
OF GOODNESS

Head
Eye
Ear
Ear Lobe
Comb
Nostril
Beak
Face
Wattles
Neck
Cape
Back
Sweed of Back
Shoulder
Cushion
Front Neck
Tail Coverts
Main Tailfeathers
Wing Front
Secondaries Wing Bay
Primaries Flights
Primary Coverts
Breast
Wing Bow
Body Feathers
Wing Coverts
Fluff
Thigh
Spur
Hock
Shank
Toe Nail
Scale

CHAPTER 3

# How Chickens Work

Grab a cup of coffee and pull up your chair. It's time for chicken anatomy class! We all strive to be the best chicken keepers that we can be, and knowing basic anatomy is an important part of that.

Chickens can't tell us (in so many words) if something hurts or if something is wrong, so it's up to us to know how our birds' bodies function and figure out problems to the best of our ability. If you call the vet to have your chicken seen, you'll need to be able to answer questions such as: "How does her crop feel? Any nasal discharge? Are combs and wattles pale? How does her vent look?"

It's also useful to know basic "chickanatomy" so you can understand and appreciate the absolutely wonderful treasure that is the chicken. So, let's dive in!

## Eyes, Ears, Combs, and Wattles

### Eyes

Chicken eyes are remarkable. They are situated on either side of the chicken's head, which allows for 300-degree vision. This is true of a lot of prey animals, including poultry, who need to evade predators. It's not quite "eyes in the back of the head," but it may be the next best thing!

Chickens have eyes in one of four colors: golden orange, copper red, brown, and black. Like humans, chickens do see in color, but they see colors better than we

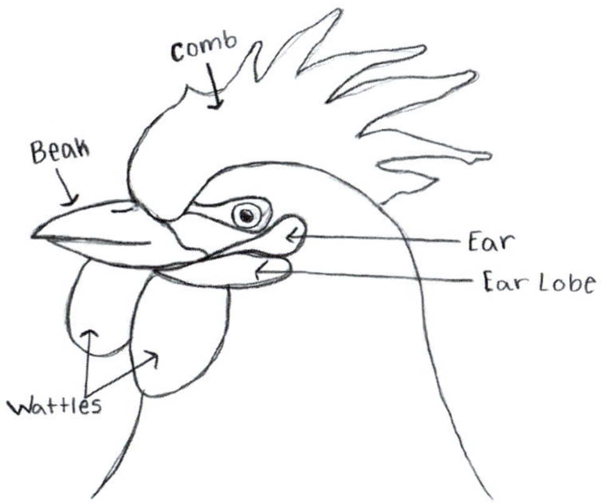


*Diagram of the chicken head.*

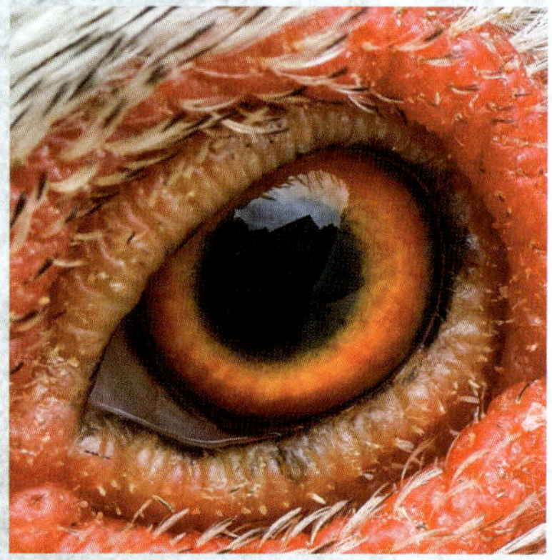

do. We have three cones in our eyes that allow us to see red, blue, and green, and variations of those shades. Chickens are tetrachromatic—they have all three cones, plus one that allows them to see ultraviolet light. This is one of the reasons that roosters will crow while it still appears dark to us—they can see the ultraviolet light from the distant sunrise—though there are plenty of other reasons why roosters crow. (For more on crowing, see chapter 4.)

In addition to color, chicken retinas have a double cone receptor that is excellent at detecting and following motion; that's what makes them such good bug catchers. In total, chickens have five different types of cones in their eyes. Scientists have found that the cones are distributed throughout the retina, but two of the same cones are never next to each other. This maximizes the effectiveness of the five different cones.

Another feature of the chicken eye is the nictitating membrane, or third eyelid. This is a transparent membrane that slides horizontally from front to back. It has a gland that lubricates the eye as well. It is designed to protect chicken eyes during dust baths and dusty foraging. Chickens blink with this membrane.

## Ears

The chicken ear is another complex and wonderful feature of chicken anatomy. Chicken ears are small round openings on the sides of their head and usually covered by feathers. While still in the egg, a chick's ear starts developing on day twelve. The earlobe is usually visible on chickens, with the exception of large-crested breeds.

Earlobes range in color from bright red lobes, to the often-large white lobes of Mediterraneans, to the blue lobes of Silkies. There is a partly true myth that claims you'll know what color eggs your hen will lay based on the color of her earlobes. In general, white egg layers do have white earlobes, but there are exceptions. The Penedesenca has white lobes and lays a dark brown egg. Most hens with red earlobes lay brown eggs, but Legbars and Araucanas lay blue and green eggs. And as much as we would love Silkies to lay bright blue eggs, they lay white or cream.

*White earlobe on an Andalusian hen.*

The avian ear structure is much simpler than that of mammals. Where we have outer, middle, and inner ears, and an "advanced" spiral cochlea, chickens have inner, middle, and outer ears and a curved cochlea that is simpler in form than the spiral. We also differ in the fact that we as mammals can hear higher-pitched sounds than chickens, while they can hear lower frequencies than we can. Scientists theorize that things like earthquakes, distant thunder, and even avalanches are audible to chickens, while they don't register with us until they become louder or the pitch changes. Chickens also have the remarkable ability to repair their own eardrums to avoid developing deafness.

## Combs and Wattles

Combs and wattles are some of the most distinguishing features on chickens, but how much do you really know about them? The comb is the red, ridged structure that sits on top of a chicken's head. Essentially, their crown. Hens usually have a smaller version, while roosters can have very large combs. Combs are red because of the melanin present in their dermis (or skin) layers. Chickens have different types of comb depending on their sex and breed. Comb types include straight, rose, pea, walnut, V, crown, strawberry, and cushion.

Wattles are flaps of red skin that hang down on either side of the chicken's jaw. Like combs, they vary in size based on sex and breed. Faverolles have very little wattle, while Jersey Giants have a lot of wattle, especially the roosters.

The primary job of combs and wattles is heat regulation. The bigger the comb and wattles, the more heat they give off. This is why in breeds from hot climates, like Leghorns, Andalusians, and White Faced Black Spanish, both hens and roosters tend

*Salmon Faverolles hen*
*with virtually no wattles.*

*Jersey Giant rooster with a*
*big comb and lots of wattle.*

to have a lot of comb and wattle. Conversely, breeds from cold climates, like Salmon Faverolles and Chanteclers, have very little comb and wattle.

The secondary job is sexual selection. Roosters shake their wattles at the hens in an effort to entice them. We're not sure if the hens really like it, but at least they know what to expect.

Combs and wattles also communicate a lot of information to chicken keepers about the health of a chicken, including sexual maturity, but also illness and overall health. A pale comb can mean serious illness or other physical problems, while a floppy comb sometimes indicates dehydration.

## Respiratory System

Birds, including chickens and other domestic poultry, have a unique respiratory system. Instead of flexible ribs that move as the lungs expand, chickens have rigid lungs and multiple air sacs that expand and contract to move air. The overall result is a body much lighter in weight. In fact, some of the air sacs are connected to hollow bird bones, and this aids in flight.

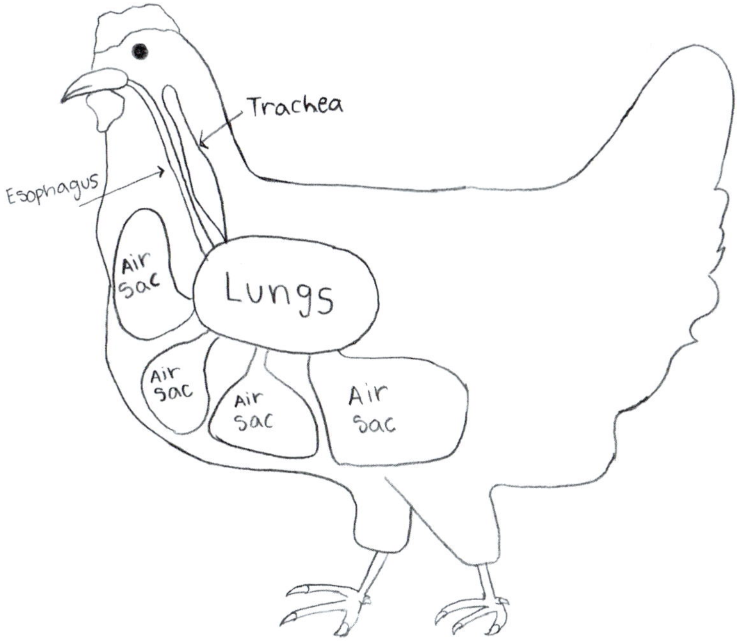

For a chicken, breathing begins with inhaling through their nostrils—or their beak, if they're suffering from certain types of stress. The breath travels down the trachea and into the bronchial tubes, leading into the lungs. Air passes into and through the lungs and continues on to the air sacs. On the way, the air passes through parabronchi in the lungs, allowing for an exchange of oxygen for carbon dioxide.

Chickens have a total of nine air sacs. There are four pairs, and one single sac. All of them expand and contract, drawing air out of the lungs and keeping it flowing

*Chickens in the Polish family are known for their style and their large nostrils and vaulted skulls.*

through the body. The air sacs connect to hollow or pneumatic bones in the chicken's body. The pneumatic bones include the skull, humerus, clavicle, keel, pelvic girdle, and lumbar and sacral vertebrae. Breaking any of the bones may cause a chicken to have difficulty breathing.

Though it is extremely efficient, a chicken's respiration can be negatively affected by dust and other particles in the air, too much ammonia, or cold drafts. Open-mouthed breathing should always be investigated. In some instances the chicken may be hot and is panting to release heat and bring down her body temperature. It is also possible that the bird has inhaled something harmful, which can damage her air sacs. This is why it is important to keep their environment clean and as free of toxins or airborne fumes as possible.

## Digestive System

The chicken's digestive system is highly efficient and quite different from mammals, and is also found in turkeys, quail, and other birds. What sets them apart is the

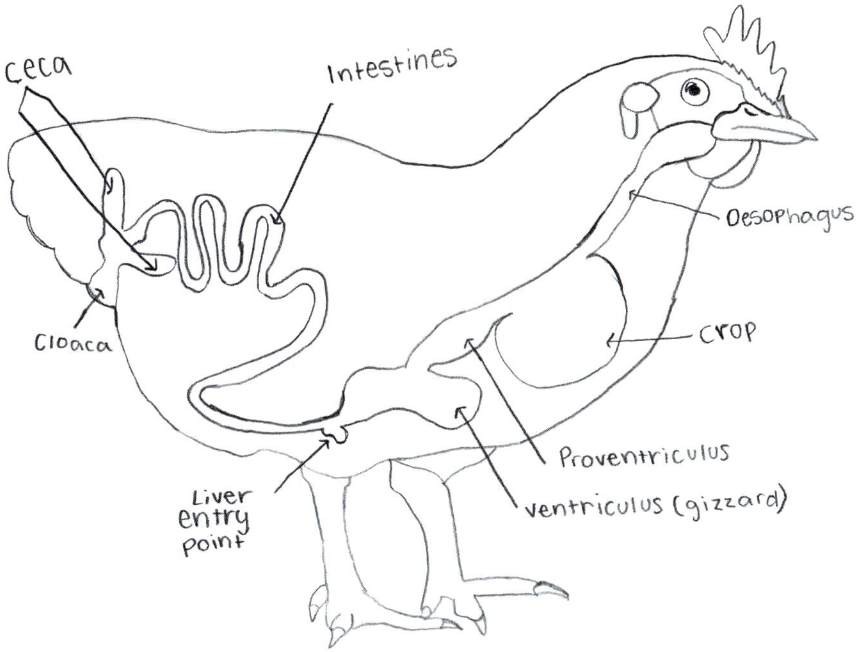

presence of the crop. The crop is an enlarged pouch located on a chicken's esophagus. It sits to the right of the trachea on the right side of the bird. It serves as an external stomach—a doggy bag—where your chicken can gobble her lunch and go. She'll digest the contents over the course of the next few hours. The crop is a holdover from your bird's Dinosaur Era, when eating on the run was a true requirement for staying alive. The crop is made of thin esophageal tissue, is fastened to the skin and breastbone with loose connective tissue, and has two muscles that support it and help it move food through the system.

As the chicken eats, the food collects and sits in the crop while the other digestive organs are full of food. Once there is space, the supportive muscles contract and push more food along the way through the digestive tract. The crop normally empties every two to four hours. Chickens pack their crop full of food during the day and it slowly digests through that day and overnight. When a chicken wakes up in the morning, her crop will be flat, or mostly flat.

The proventriculus, also known as the glandular stomach, receives food from the crop. The proventriculus secretes digestive fluids and enzymes to help digestion. It is the second stop for food on the digestive train.

The gizzard, or ventriculus, is the third stop in the digestive system. Chickens and other birds aren't the only ones with gizzards—dinosaurs had them, of course, but alligators, crocodiles, certain fish, some mollusks, and some worms do, too. The gizzard is a round and hollow structure, basically a very strong/tough bit of muscle that takes the place of teeth in poultry and all of the other critters we mentioned. Chickens will pick up grit, sand, and tiny pebbles to help the gizzard do the grinding. This is why we advise you to have commercial grit available at all times in your chickens' runs.

Food that has been ground in the gizzard moves on to the duodenum, the start of the small intestine. It has three parts: the duodenum, jejunum, and ileum. This is where food is digested and nutrients are absorbed. The intestine is also equipped with several features that help with the chicken's overall immunity.

Coarse bits of food or plant debris that aren't digested in the small intestine are filtered into the cecal pouches. These pouches, which are located where the small intestine ends and the large intestine begins, are equipped with lots of beneficial bacteria that help ferment and break down fibrous material while absorbing water and more nutrients.

The remaining material from the cecal pouches passes into the large intestine. More water is absorbed here, and the remaining undigested material is condensed. From here, the condensed matter—better known as chicken poop—moves into the cloaca to mix with urates.

Urates are filtered by chickens' kidneys. They are crystallized into uric acid and secreted into the cloaca (vent, or chicken butt). Chickens don't produce liquid urine; the urates are the white "cap" that you usually see on chicken poop. This final mix is passed through the vent.

*Bluebelle the Swedish Flower hen, sporting a well-filled crop.*

## Reproductive System

In chapter 2, we went over the journey of an egg through a hen's reproductive system. Now we're going to tell you more about the reproductive system itself.

Hens manage to produce hundreds of eggs for us, and they do it all with just one ovary. It is worth paying close attention to your hens' reproductive health, because this is one of the places where problems can occur. In chapter 9 we will discuss reproductive problems in more detail and share some best practices for your hens' health.

The ovary sits about halfway down a hen's back. When she hatches, her ovary is fully formed and holds hundreds and hundreds of tiny follicles containing tiny

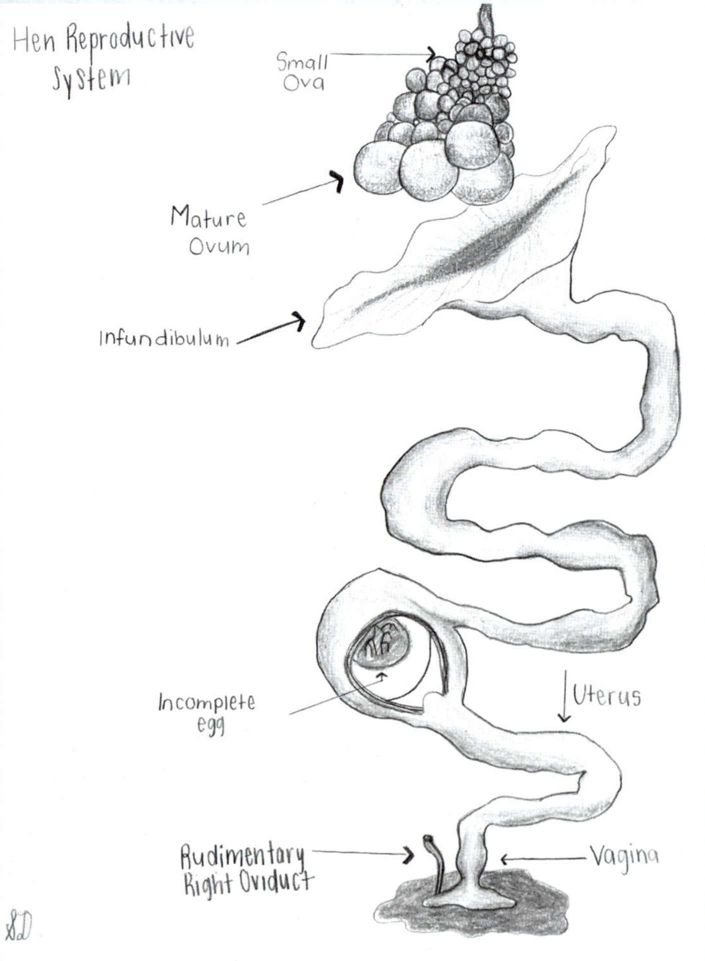

*Margo, a Buff Orpington, has laid her egg and is coming out of the coop.*

ova. These are all the eggs she will be able to produce in her lifetime. When a pullet reaches sexual maturity, she will begin to ovulate, or release an egg to enter the oviduct and start the formation process. The ova will start to grow a few at a time, and the ovary will release a single developed ovum roughly every twenty-four hours or so. The ovum enters the oviduct and the egg formation begins. The egg passes out of the oviduct, into the cloaca, and out through the vent.

Ovulation and egg laying are tied to day length. Daylight stimulates the hen's pituitary gland to secrete luteinizing hormones. This causes her to ovulate. When daylight starts to diminish, the hormone release is slowed, or stopped. She will not produce eggs again until there is enough daylight to stimulate her pituitary gland.

The male reproductive system is straightforward and efficient. Roosters have a pair of testes located about halfway down their back. When sperm is produced, it travels via tube to the rooster's cloaca. When chickens mate, the rooster must press his vent against the hen's vent and release sperm. He does not have a penis and there is no penetration. Any sperm that enters the hen's oviduct is stored in the infundibulum. When the egg passes into this part of the oviduct, any sperm present can perforate and fertilize the egg.

## Skin and Feathers (Integumentary System)

The integumentary system of the chicken is fascinating. The entire system is composed of skin, feathers, claws, beak, and glands. We're mostly going to be concentrating on the skin, but all of the system provides important functions that are a source of rich information about your chicken's health. The integumentary system also protects your chicken from the elements, allows them to sense danger and changes in the atmosphere, insulates them and controls feather replacing through molting, and lets them do all of the things that chickens need to do.

Chickens have two layers of skin: a thin dermis inside, and a thicker epidermis forming the outer layer. Both levels are mostly made up of collagen, elastin, and keratin. There is also a sublayer beneath the dermis called the hypodermis. This is the area where lipids and fat cells are stored.

The thicker epidermis is where a lot of the action happens. This outer skin contains feather follicles that generate new feather growth during a molt or after injury. The feather emerges from the skin via the feather follicle with nerve and blood supply

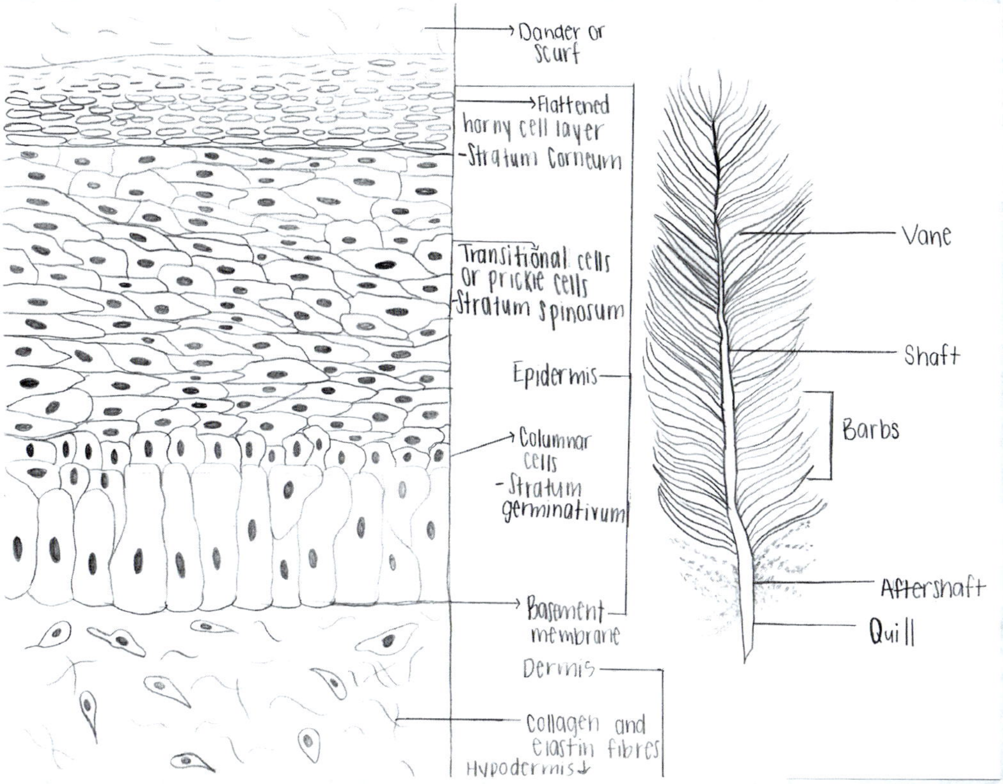


*Skin and feather follicle.*

intact and connected to the dermis. The blood supply will dry back up once the feather has fully emerged and unfurled. To us it may look like feathers are just popping out willy-nilly, but the follicles are actually attached to precise tracts of skin—and different tracts may grow different types of feather.

Chickens molt and regrow their feathers once a year, usually in late summer or fall, and it requires a lot of their resources to do it. Molting chickens need a low-stress environment and good nutritional support to help them grow the hundreds of healthy new feathers that they'll require to keep warm over the winter.

Unlike mammals, avian skin does not contain sweat glands or sebaceous glands. Chickens don't sweat, and their skin does not excrete oil. Chickens groom themselves by dust-bathing and preening, which involves collecting oil from the uropygial (or preen) gland and rubbing it onto their feathers. The gland is situated on their back, near the base of the tail, and it has two lobes. The gland secretes oil that the bird

Silkies do have the preen gland, but because their feathers have no barbules, they cannot waterproof themselves, even if they tried.

collects on her beak, and she spreads it along the feathers to groom and condition them. Preen oil is generally made up of waxes, fatty acids, and volatile organic compounds. Dust-bathing removes excess oils, skin and feather dander, and helps combat mites and lice.

When a chicken or any other bird preens, they are using their beak or bill to condition their feathers. This may involve smoothing the feathers so that the barbules catch and stay together and improve weather resistance. Preening also involves placing feathers that may have become disarranged or ruffled back into their correct position; picking out mites, lice, and anything else that doesn't belong in or on feathers is also an important part of the process.

While most of a chicken's body is covered with skin beneath their feathers, the skin on their feet and legs is mostly keratin formed into scales. The scales are attached to the skin of the epidermis and carry all the way down a chicken's lower leg to the top of their toes. The bottom of the foot has thicker skin that has evolved to withstand pressure from the bird's weight, and to minimize abrasions and injury.

*Feathers from Spicy, the Rhode Island Red.*

The beak and toenails are also primarily composed of keratin and have both blood vessels and nerves. They allow your chickens to explore their world, eat, bathe, defend themselves, etc. Broken beaks and toenails will need first aid if severe enough. With the exception of severe damage, beaks and nails will grow back over time.

## Chicken Brain

We like to refer to the chicken brain as a microchip. The brain, which sits toward the back of the skull, is tiny—about the size of a walnut—but extremely precise.

Like the human brain, the chicken brain is divided into two hemispheres. These hemispheres work independently, each performing different tasks. A neuroscientist named Leslie Rogers pioneered this area of research on chicken brains. Unlike the human brain, chickens can use the two sides of their brain to carry out different tasks at the same time. For example, a chicken can use her right brain to look for food while her left brain is scanning for predators. The chicken can also go from seeing an image with both eyes—binocular vision—to seeing two separate images with no overlap—monocular vision.

The chicken's left brain is responsible for detailed work like finding food, while the right brain takes care of fear and escape responses. The sides of the brain work with opposite sides of the body. The left side of the body is linked to the right side of the brain, and the right body is linked to the left brain.

*Egyptian Fayoumi, smart and beautiful!*

The chicken's brain is also responsible for their excellent memory and inter-flock social skills. Scientists have proven that chickens can remember up to a hundred faces, both animal and human. They also possess cognitive reasoning ability that allows them to perform simple math, estimate time intervals, and recognize object

In 2009, researchers in the psychology department at the University of Padova in Italy used a variety of tests, including moving objects and hiding treats, to successfully prove that newly hatched chicks had the ability to count, add, and subtract numbers smaller than 5.

permanence. Chickens are also highly trainable, including housebreaking!

Scientists are also studying the emotional range of chickens; so far they have identified fear, joy, and even empathy. In our own flocks we have seen sadness/mourning. Chickens' emotions manifest as vocalizations, body language, and their social interactions with flock mates and their keepers. There's no mistaking the happy clucks from hens about to get their favorite treat or a contented chicken purring in your lap as she sleeps.

## Normal Chicken Behavior

It may seem odd to end a chapter about anatomy with everyday chicken behaviors. But, for a new chicken keeper, knowing what "normal" looks like can really help with peace of mind. While normal varies from chicken to chicken and flock to flock, there should be some basic behaviors that they all have in common. We'll also discuss this more in the next chapter.

Eating, drinking, and pooping are the basics. Chickens will eat and drink throughout the day. Eating can include picking from the food dish, scratching for bugs, eating scratch grains or treats, and eating grass and other plants. Chickens will visit their waterer frequently, and that is completely normal. It is also normal for your chickens to poop. A lot. A healthy chicken generally poops ten to fifteen times per day.

Social behaviors are also good examples of chickens behaving normally. Hens will stand in a group and preen themselves, and sometimes groom each other. They will lie down near each other and nap, often in the afternoon. Some birds prefer to perch during the day and rest that way. You'll also see your birds dust-bathing and sunbathing; these are both normal and healthy activities.

It's a good idea to spend time watching your flock going about their business every day. You'll get to know their habits and patterns, making it easier to spot problems much more quickly.

*Lavender Orpington hens roosting together.*

CHAPTER 4

# Planning Your Flock

We're hoping that by now you have decided to get yourself some chickens. Not only is this the best decision ever, but it will pay you back in joy—and eggs—for a long time to come. In order to make sure your path to the coop is clear, so to speak, it's a good idea to take some time and do some in-depth research before you run out and get your birds.

Most of the things you want to think about are practical, and you may be able to change them later. But it is always easier to think things through and try to solve as many problems as possible before you have your chickens. In this chapter we're going to tell you about some of the more important considerations you'll face when getting started. This is the poultry equivalent of measure twice, cut once.

## Laws, Ordinances, Zoning, and Regulations

We would love it if anyone who wanted chickens were able to keep them, but that is not reality at this point in time. While there are no federal laws regulating chicken keeping, and few state laws, the majority of regulations you'll find are county and local level.

It is important—and we cannot stress this enough—to make sure your town, city, village, neighborhood, parish, borough, township, or municipality allows you to keep chickens. There is nothing worse than getting set up and falling in love with your girls and then having to re-home them because you have inadvertently broken a law.

The laws vary. Many areas do allow chickens, but they may not allow roosters, or they may limit you to four or six hens, or you may need a license or permit to keep any poultry. One of our local counties requires you to submit coop and run plans for approval. We know that some people keep chickens on the DL, but that is a recipe for disaster. You could be fined, and even worse, you could lose your chickens.

The first place to look for chicken-keeping laws is your county code. In many areas, the county laws are available online and you can find them with a basic Google search. If you live in an area where county laws haven't been digitized, you can often find bound versions of the laws in your local library. Failing that, try your town or county courthouse.

There is one more entity that has the power to stop you from keeping chickens: a homeowners association. Once you have signed the covenant papers, there is often little you can do to get the rules changed. We want you to be aware that HOAs are a formidable obstacle, but there *have* been legal cases where covenants were changed to allow chicken keeping.

If you decide to work at getting local zoning laws and regulations changed, there are some resources that may help. Many neighborhoods and towns have active

Facebook pages where they share information about ways to approach local lawmakers. These groups are usually generous with their information and support. You can also start your own group and try to find like-minded chicken people in your area to join the fight.

## Think Outside of the (Nest) Box

Not everyone is lucky enough to have a traditional backyard. But, with a little creativity, you may still be able to keep chickens. You want to make sure that the chickens will be safe and secure, that they have adequate space to move around, and that their coops and runs (fenced exterior sections that may be freestanding or attached to the coop) are strong enough to withstand bad weather and protect them from the elements.

City dwellers often think that they cannot be chicken keepers, but cities like New York, Baltimore, Chicago, and Los Angeles do allow chicken keeping. But what if you don't have a yard? You may want to consider rooftop chicken keeping. If your local ordinances and your building allow it, you can set up a coop and run in a safe spot on your roof. You'll want to use strong materials to withstand wind and weather, and make sure the run has plenty of space. It's generally not safe to free-range your chickens on a rooftop.

Another option for apartment dwellers is keeping chickens in community garden spaces. If you're lucky enough to have a community garden, check with the manager about the possibility of setting up a coop and run space. In cities like Baltimore, community gardens are a vibrant part of neighborhoods. Residents can gain food security by growing food and keeping a flock of laying hens.

City dwellers who do not have community gardens can look into starting one. For instance, Baltimore City has a program where residents, businesses, or neighborhood groups can sign up for the Adopt-A-Lot program. The group adopts a vacant lot that they can turn into a community garden or food-growing space. Bigger lots have the opportunity to keep a larger number of chickens by permit.

Another unusual solution to keeping chickens is gaining momentum: house chickens! Before you either laugh or turn up your nose in disgust, there are plenty of people out there who make this work. Sometimes the chicken is elderly or has special needs, and will be safer and better cared for in the house. Or, the chicken(s) may be so beloved

that their people want them around all of the time. There are lots of options on the market for chicken diapers and liners, and chickens really do make great snuggle buddies.

## Basic Flock Behavior

If you are planning to get chickens and you've never been around them for any length of time, there are some behaviors that you will want to be familiar with. Chickens, while very intelligent, are creatures of habit, and we think it is wise to be forewarned about some of their quirkier ways.

*Eclair, a young Light Brahma, enjoying a sunbath.*

Chickens love dirt. If you keep them in a place without access to dirt, they will clear the ground and make dirt. Few animals will defoliate a patch of ground faster than chickens. They love to scratch and dig and roll in the dirt. This is part of their daily life, and the entire flock will engage in this behavior.

Dirt allows chickens to dust-bathe, as we discussed in chapter 3. Chickens will also sunbathe, and it looks a little scary the first time you see it. The chickens will be lying on their side with a wing stretched open, often with one leg sticking out and their hackle feathers up. They're fine. In fact, they're probably very relaxed and happy, and they won't appreciate it when you rush up to make sure they're still alive.

Chickens communicate with each other constantly, and much of it is vocal. Roosters crow, yes, but chickens have over thirty distinct vocalizations. Hens will sing a song after they lay an egg, various flock members may give an alarm call if they see something that worries them, and roosters and some hens will call others over to eat something especially delicious. This last vocalization is known as tidbitting.

It often surprises new chicken keepers to learn that hens prefer laying in one or two nest boxes. The coop could have eight boxes, but only two or three will contain eggs.

*Basil, an Andalusian hen, heading for her favorite nest box.*

Sometimes the hens will even fight over those boxes. It's frustrating for some keepers, but there is a simple explanation. Hens have a biological imperative to lay a clutch of eggs. They all lay in the same few boxes because they know that the eggs are likely safe there, and because the odds are better that a broody hen may choose to sit on a bigger clutch of eggs. This is very important, because the broody hen will hatch and raise the next generation and ensure the survival of their species. Broodies will sit on any eggs that will fit under them, including eggs from other breeds and even other species, like ducks. You can use false eggs to help with some of this nest-box drama. If you put a few in unused boxes, the hens may think they look like another good place to lay.

## Pecking Order

The term "pecking order" is straight out of the chicken-keeping world. We don't always think of pecking order as a good thing, but it is an important feature in a flock. Pecking order evolved so that each chicken knows their place and the flock has a certain hierarchy to it. This doesn't mean that the order never changes, but it does mean there is always a leader or two, and there are always a few hens at the bottom of the ladder. The head hen will eat first, and she will have first choice of nest box and roosting bar. She may settle disputes and she may decide where the flock free-ranges. In a well-balanced flock, she may not do much.

Next in line are the birds that we call the deputies. There may be one or two of them, and they may spend a bit more time than the head hen regulating the flock. The deputies are quick to put a lower-ranking hen in her place when they feel it is necessary. The rest of the flock know their place and usually exist in harmony unless something has happened to upset the pecking order and cause them all to jockey for position.

Anytime there is a change in the flock, the pecking order resets. This includes flock deaths, new chickens being added or removed, severe molting, or removing a chicken from the flock to treat illness. In general, you can let the pecking order sort itself out. The only time we intervene in pecking order squabbles is if a youngster is being bullied and has nowhere to go, if a bird is severely stressed, or if someone is injured or bleeding. You should never let birds fight until there is bloodshed or trauma.

You may think that starting your flock with only one breed will do away with pecking order trouble, but it will not. Even in a same-breed flock, a pecking order will be established—although the jockeying for position can be less intense.

*Sweet Cochins are usually near the bottom of the pecking order.*

Pecking order is heavily influenced by the breeds of chicken that you keep in your flock. More assertive chicken breeds, or those with strong personalities, tend to be higher than gentle breeds. For example, breeds that tend to be high in the pecking order are Marans, Wyandottes, Rhode Island Reds, Andalusians, Plymouth Rocks, and Swedish Flowers. Gentle breeds that tend to be at the bottom of the pecking order include Salmon Faverolles, Brahmas, Cochins, Polish, Delawares, and Orpingtons.

It is important to note that a breed may have a strong personality within the flock, but they can still be very docile and gentle with humans. It is also worth noting that age and seniority count for a lot in the pecking order. For years, two Orpington hens were the queens of Chrisie's flock. They were her original hens, and they stayed in charge for ten years before they passed away. They were kind leaders and kept the flock balanced and peaceful.

## Roosters and Rooster Plans

We love roosters. They're beautiful, colorful, and full of love! Plus, the crow makes us happy every time we hear it. If you don't like the crowing, or if you aren't allowed to have roosters, don't worry. You can be a very successful Chicken Lady or Gent without ever keeping a rooster.

In our opinion, roosters get a bad rap. They are very good at taking care of their hens, keeping an eye on the sky for predators, and enticing their ladies with the best snacks. Some roosters can be overprotective, but most boys respond very well if you work with them. An overprotective rooster is doing the job that he is hardwired to do: defend his hens.

If you just want pet chickens and aren't worried about having hens for eggs, we encourage you to get roosters. A rooster who doesn't have ladies to look after is one of the best pets ever. They're sweet and affectionate, and get along well with other roosters and pets as long as there are no hens within sight or hearing. We know people who swear that their evening cuddle with a rooster in front of the TV is the best part of their day. It isn't hard to find roosters that need homes. Shelters and humane societies are full of them.

*Kasper, a Swedish Flower Hen rooster.*

**Pros of keeping roosters**

Protect and defend the hens

Fertilized eggs, making them viable for hatching

Keep peace in the flock, separating squabbling hens

Beautiful to look at, gorgeous form and feathers

Fewer health problems than hens

Eat less than hens

**Cons of keeping roosters**

Overmating with hens, causing feather loss and stress

Loud crowing, often at all hours

Sometimes aggressive with humans

Accidental injury of hens during mating

No eggs

Large combs and wattles require protection from frostbite (see chapter 9 for information on frostbite prevention)

*Little Man, a Polish rooster.*

If you have a rooster, you'll quickly notice that most of his time and focus is on his hens. He makes sure they're protected, shows them where to eat and drink, and encourages them to use the nest boxes—sometimes even getting in the box himself to show the girls that it's safe. All of these behaviors are part of the rooster courtship ritual. There's also a special dance that a roo will perform for his ladies. It's called the wing dance, and looks like a little shuffle, with his wing extended down to his foot. You'll know it when you see it. Most roosters dance before the act, but sometimes youngsters are so eager that they forget and give a quick dance right after. Better late than never!

A rooster will breed with every hen in his flock. If he doesn't, it's because the hen has somehow managed to avoid him. He may mate with each hen more than once, especially if he's young and full of testosterone. And at some point, he may decide that you or someone else in your family is a danger to his hens. He may try to attack you by biting, or, more commonly, spurring you. A rooster's spurs are located on the back of his legs and can be quite sharp.

If your rooster starts displaying aggressive behavior, you can work with him if you choose. You can also give him a time-out away from his ladies, and you can spend time with him one on one. Carry him around, snuggle with him, even try teaching him basic things like coming to you when his name is called. Back in the run, you can hand-feed him and the hens, and take note of what you're doing when he begins to show aggression. Holly's Leghorn rooster, Giorgio Armani, had begun attacking when she went into the run to give out treats. After paying close attention, she realized that he was going after the treat bag, not her. She started dispensing treats from a bucket and Armani settled down.

*Rooster spurs and nails can grow long and need occasional trimming.*

In our experience, roosters hit the peak of their testosterone at about eighteen months. This is when they mate the most and when they tend to develop any aggressive

*Katsumoto, an Australorp rooster.*

behaviors. As we mentioned, roosters almost never show any aggression if there are no hens that they can see or hear. They are all different, and have their own quirks. Chrisie's Polish rooster, Little Man, is typically very docile, but gets upset when he sees anyone collecting eggs. Holly has twelve roosters, and for the most part they're all very laid-back.

If you absolutely cannot or do not want to have a rooster, we have a few suggestions. We encourage all chicken keepers to have what we call a Rooster Plan. This is knowing ahead of time what you will do if you end up with males when you purchase or hatch chicks. Some of our homesteading friends choose to grow out the young cockerels for a year and then use them as table birds. Another option would be to re-home the boys, but exercise great caution when doing this. Some people want to use free roosters for inhumane purposes. Other keepers may choose to keep a bachelor flock where it's all boys, and not a girl to be seen. Half of Holly's roosters are Nankin bantams living in a bachelor flock. They are very sweet and affectionate and she loves spending time with them.

*Holly's bachelor flock, supervised by Daphne, a Jacob ewe.*

*Beatrix Potter, a Legbar hen.*

*Little Jackie, a Bielefelder hen.*

Whether you are buying chicks at a feed store or ordering from a hatchery, make sure you select sexed pullets and not straight run chicks. "Straight run" means unsexed. "Pullet" is the term for a young female chick. The best option for someone who needs to avoid roosters altogether is to purchase auto-sexing chicks. Auto-sexing essentially means that the chicks are sexually dimorphic; males and females have different colors and markings at hatch. Two of our favorite auto-sexing heritage-breed chicks are Crested Cream Legbars and Bielefelders. They're both great for beginners, and we'll tell you all about them in chapter 6.

## Heritage Breeds

Heritage breeds are livestock and poultry breeds that have been traditionally used in a certain geographic area, usually over several generations. For example, the Rhode Island Red chicken was developed in New England and has been kept and used in the United States for more than one hundred years. Heritage breeds are pure-breed chickens that have been developed to look and behave in certain ways. They also breed true; breeding two birds from one breed together will produce babies that look and act like their parents. Two Rhode Island Reds gives you baby Rhode Island Reds.

One of the most important things that heritage breeds offer is their biodiversity. Each different heritage breed comes with its own complex genetic makeup. Their unique traits may include things like egg-laying ability, heat or cold tolerance, disease resistance, appearance, and even personality.

Heritage breeds connect us to history. When you keep heritage-breed chickens, you are part of a larger tradition of ancestors who kept chickens as part of daily life. This tradition spans thousands of years, but it started to change after World War II. As people moved away from rural living, chicken and egg farming became industrialized, and hybrid laying hens were developed. Heritage breeds did not fit into this new commercial mold. Many of the breeds began to slip toward extinction.

Hybrid laying hens have been developed to produce large quantities of eggs, more than most heritage-breed birds would produce. Some hybrid hens will lay well over three hundred eggs per year. Hybrid chickens are varieties created by crossing particular breed stock. The parent breeds are always the same, and the hybrid offspring will not breed true. Most hybrids were developed by genetics companies to create superlayers for industrial egg production.

Hybrids tend to be very docile and gentle birds. The downside of hybrids is that they have much shorter lives than most heritage-breed birds. Because they have been engineered for mass quantities of eggs, they often develop reproductive diseases and cancer. There are many hybrid layers available to purchase from farm supply stores. They can make a great addition to a backyard flock, but be aware of the potential for increased health problems and higher mortality rates. Because hybrids are so gentle, also be wary of adding them to a flock of chickens with strong personalities.

Most heritage-breed poultry have breed standards accepted by the American Poultry Association. These standards are found in the Poultry Association's *Standard of Perfection*, a bound book that is updated every few years. The different breeds are divided by class, and most chickens are classified as American, English, Asiatic, Continental, Mediterranean, or All Other Standard Breeds. If you're unsure which breeds mix together well, one secret is to choose all breeds from the same class. There are some exceptions, of course, but in general, all American breeds do pretty well together—all Mediterranean, all Asiatic, and so on. As you become a more experienced chicken keeper, you'll know better what you like to have in your flock, but the classifications are still useful as a way to generally understand the overall traits of groups of chickens.

*Spicy, a Rhode Island Red.*

CHAPTER 5

# Where Will Your Chickens Live?

Now we're getting to some of the fun stuff: chandeliers and fairy lights, nest-box curtains and seasonal wreaths. You've got Pinterest boards full of ideas, and you know exactly which bedding you want to use, and why. Or maybe you just want a solid coop that fits in with your aesthetic and you're not worried about making it IG-ready. Perhaps you're at the point where you want chickens but you're overwhelmed at the choices and don't know where to start. The Chicken Ladies are here to help!

It is so much fun to decorate your coops and runs, but none of that matters if you don't have the right setup. Coops and runs are your first line of defense against predators and the weather. And if you get it wrong, you can also cause a lot of physical discomfort and mental anguish for both you and your chickens. Nothing is worse than a setup that makes more work for you, so if at all possible, do plenty of research and planning before you jump in with both feet.

## Types of Coops

You can find all kinds of coops out there if you look around long enough. We've seen anything from repurposed sheds to old cars and buses to cob/clay coops. But the two most common types of chicken coop are wood and plastic. Wood coops have been in use for thousands of years, while plastic coops are obviously much newer. Both have pros and cons, but you can make either one work for you in just about all situations.

The biggest coops tend to be constructed from wood. Wood lends itself to larger projects, since it is easy to find or buy and transport, and can be cut and assembled in many ways. If you are limited in space or just want a smaller coop, wood can do

the job just as well. A coop made of good-quality wood, solid and well-built, does an excellent job of keeping the weather out and your chickens safely inside. However, wood does require maintenance. It can rot if exposed to harsh weather, and it can also be damaged by animals chewing it, or even broken by falling storm debris.

Plastic coops usually come unassembled as a kit. Most are relatively quick and easy to put together, making them accessible to people who cannot or do not want to

build with wood. Because plastic is either molded or cut to size, the coops tend to be smaller and not very easy to customize. However, plastic coop companies usually have drawings and diagrams of their coops available—an important detail for chicken keepers who need to submit coop plans for a permit. Plastic coops also tend to work very well in small spaces, so they're popular with chicken keepers with limited space. One advantage that plastic has over wood is its insulating properties. We'll share some very effective ways to take advantage of this later in the chapter.

## Building or Buying

If you've been debating buying a coop versus building a coop, price will have a big effect on the type that you get. You can't do much building with plastic, so building a coop usually means you'll be working with wood. This can be expensive if you are going for good quality. Expect to spend between $1,000 and $3,000 (and upwards) for a small to moderately sized well-built wooden coop. Your money will go a lot farther if you build, though the price of lumber and supplies and delivery fees have increased substantially over the past few years.

*Holly built a simple wooden A-frame coop for her second flock.*

Plastic coops, because they're smaller and usually unassembled, tend to be less expensive than wooden coops. As of this writing, the smallest plastic coops on the market are around $500 to $800, and the largest and most expensive range between $1,300 and $2,600. Plastic coops do have the advantage of being very easy to clean, both inside and out, and they're easier to move around, and even disassemble, if you'd like, for moving and storage.

We've both assembled a variety of plastic coops over the years, and one of the things we like about them is the fact that pretty much anyone can put them together and get themselves some chickens. You can also sometimes find both plastic and wood coops secondhand. Just inspect them very carefully for damage and give them a thorough deep cleaning before putting birds in them.

## How Big Should It Be?

Ask almost any chicken person out there how big your new coop should be, and they'll tell you to buy the biggest coop you can afford. This idea does have merit, especially if you think you'll be getting more chickens. But bigger is not always better. Take a look at your flock goals and be honest about where you think your chicken keeping may go over the next five years. If you know you're not going to be moving house, and you don't want to spend tons of time cleaning coops, start off with a coop that fits a modest number of hens and has a bit of extra space for a couple of extra girls or a boy in the future.

## Ventilation, Windows, and Doors

Whether you're designing, building, or buying your coop, you want to make sure there is good ventilation, and that doors and windows are strong and functional. Predators will try all entry points on a coop, so it's important to ensure that not only are doors and windows well-constructed, but also that there's a way to latch or lock them. Carabiners are great for securely locking latches. Windows are generally safe as long as they're shut. But when they're open during warm weather, you may need to nail hardware cloth over the top of the window frames so that nothing can tear open the window screens.

Ventilation is extremely important for the health of your chickens. Chicken droppings are high in ammonia, and ventilation is necessary to release the fumes. Windows do count as ventilation, but ideally you will have vents or some other kind of secure opening at the top of your coop. You need working vents all year long, especially when your coop is closed up for winter weather. Most plastic coops come equipped with vents. You can retrofit vents into a wooden coop if you're handy with a power drill.

While ventilation is very important, you'll also want to make sure that your coop is not drafty. A cold draft blowing onto chickens makes it hard for them to regulate their body temperature, and can lead to cold stress and hypothermia. Drafty coops also mean cracks and crevices where snow and rain may be able to enter, and good starting points for rodents who want to chew their way inside to your birds.

## Roosts and Perches

When it's time for chickens to sleep, they will seek out high roosts or any other handy high spot around. Chickens prefer to roost at night because they're safer from predators. But they are not perching birds in the same way that songbirds are. Chickens need wide or flat surfaces to roost on. They lie on top of their feet, helping to keep them warm, while songbirds have feet meant to clamp around a branch.

The best roosts for chickens are wooden 2x4s, placed so the wide side is flat; 2x2s are more common in plastic coops. You may even find that large-breed chickens prefer to sleep in the bedding in plastic coops. Metal is not appropriate material for roosts because it holds heat and cold and could easily burn or give your chicken's feet frostbite.

Whatever roosts you are working with, you want to clean them periodically by scrubbing with a 1:1 mixture of vinegar and water. Make sure there are no jagged bits or splinters that can cut your chickens' feet and lead to bumblefoot, a type of bacterial (or occasionally fungal) infection. Wooden roosts can be sanded or covered with contact paper for extra protection.

## Ramps and Ladders

Chicken coops are usually elevated, so a ramp or ladder is common to help your birds climb into their house. (There's almost always one hen who insists on flying in and out, and that's okay. She's special.) Some chickens do better with ladders and ramps than others. You can help by giving them lots of grip. Textured rubber matting or small strips of wood fastened horizontally across the ramp work well. Some plastic coops have ladders that both really big and really small chickens struggle with. Try putting a wooden ramp over the ladder for better access. Ramps and ladders accumulate a lot of poop and mud. They require periodic cleaning as well.

## Nest Boxes

Nest boxes are the spaces where hens go to lay their eggs. People started supplying hens with nest boxes to convince them that these are safe and private places to lay their eggs. This makes egg collecting much easier for all of us!

If you're converting a shed or other structure into a coop, you'll have to add your own nest boxes. There are lots of possibilities here. We've seen everything from repurposed buckets to milk crates. A little time on Pinterest will get you feeling creative and give you lots of ideas. If you're not the DIY type, you can buy nest boxes, both attachable and freestanding styles. Some nest boxes can even be accessed from outside of the run, making egg collecting safe and fun for small children.

We recommend one nest box for every four hens, and try to make them as roomy as possible. Larger breeds like Cochins and Orpingtons have an easier time laying in larger boxes, and sometimes you end up with two or more hens in a box. Larger boxes help prevent smothering deaths and crushed eggs.

## Bedding Options

No matter what kind of coop you decide on, you will need to provide bedding for your birds. Chicken droppings have a high moisture content, and bedding helps absorb both liquid and odors. Bedding also acts as a cushion for your girls as they hop down from the roosts in the morning. A deep layer of bedding protects feet and legs from injuries, and cuts down on abrasions on the soles of the feet that could become infected and lead to bumblefoot. Bedding also helps insulate your coop during the coldest months of the year.

The most popular bedding materials for chickens are wood shavings/sawdust, hemp, straw, and sand. We prefer soft wood shavings in our coops, but we do use hemp occasionally. Wood shavings are economical, usually less than $10 per bag, while hemp is more expensive, but lasts longer. Some folks swear by sand, but neither one of us likes it in our coops. Straw is also popular, but we have both had hens eat the straw, develop a blocked crop, and end up having surgery. Do some experimenting and see which bedding works best for you and your flock. You will want to avoid cedar chips and anything scented with cedar; it can irritate and damage a chicken's respiratory system.

We each clean our coops about every seven to ten days and replace the bedding completely. You can compost what you take out of the coop. You can also rake used bedding into your run and let it break down further as substrate. Straw, hemp, and

Pine Shavings

Hemp Bedding

shavings can all be composted, or used for the deep litter method—keeping a thick layer of bedding that is turned and added to regularly, to encourage microbial action and breakdown of organic matter—but sand cannot.

## Coop Fans and Heaters

We believe strongly that technology can and should be used to better the lives of our feathered darlings. The new generation of coop heaters includes safe options that are wonderful for preventing frostbite and keeping senior hens comfortable in frigid weather. They are thin panels that give off radiant heat and are safe to come in contact with coop bedding.

We also use coop fans during the steamy months of summer here in Maryland. It is true that chickens don't sweat, but fans keep the air moving and prevent hot air from building up in your coop. The air circulation also helps pull heat away from combs and wattles, helping chickens cool down. If you don't have access to electricity, we recommend investing in rechargeable battery fans.

Large, healthy, fluffy chickens with lots of feathers may need help during hot weather, but they generally withstand the cold well in a coop with the appropriate number of chickens. Birds that may need some supplemental heat in winter include roosters with large combs and wattles, ill or elderly chickens, bantams and small-bodied birds, Mediterranean-class chickens, and hens in a severe molt.

## Chicken Run

We are not exaggerating when we say that your coop and run can make or break your chicken-keeping experience. Your coop is like your chickens' bedroom, while the run is the rest of their house. If your run is too small and your chickens are overcrowded, you could end up with boredom, pecking injuries, and stressed, unhappy chickens. A large run, at least 10 square feet per bird, gives your chickens plenty of space to scratch and dig and even run around if it's spacious enough. Plus, a really big run can often fit a second coop if you decide that you simply must acquire more heritage-breed beauties!

Just as important as the size of your run is where you decide to put it. Siting your run requires careful consideration, and some observation as well. You'll want to take notes on how many hours of sun and shade your site gets (you will want shade if your summers are warm), the direction from which the wind usually blows, how well water drains after rain or melting snow, if it is in a floodplain, whether or not you can have electricity, how accessible it is from your home, whether or not you can have a coop or materials delivered, and any other details that will affect you and your flock as you go about your everyday business. Look at your flock goals and remind yourself what you want out of this experience; don't take shortcuts here. This is a place where return on investment counts.

Once you have your site picked out, you can start looking into whether you should build a run or buy one. Building a run, like building a coop, gets you the most size for your money. But if you can't build and can't hire someone to do it, putting together a purchased run may be a good option for you. Some of the plastic

*Holly's duck run frame.*

coop companies do offer both walk-in and attached runs made from steel-mesh components. You can also buy steel-framed runs that are a lot like dog kennels. The framing for these runs tends to be wonderfully sturdy, but the wire is often flimsy and unlikely to keep out predators. Invest in a few rolls of good-quality welded wire or hardware cloth if you can afford it, and use that on the frame instead. With a bit of modification, you can have a very strong run.

If you choose to build your run, the most common options are wood post/frame and wire, or metal T-post and wire. Wood will look great, but it's costly; T-posts are less expensive, but more utilitarian-looking. While you can cut wood posts and beam to size, the length of T-posts is fixed. For 6-foot-high run walls, we buy 8-foot T-posts and drive them 2 feet into the ground. You can attach wire to T-posts using metal clips that generally come with the posts, or with heavy-duty zip ties. Zip ties will break under heavy pressure and they become brittle in heat and cold. If you use them, you'll want to use a lot of them for added strength. Inspect and replace them often. Deck screw and U nails are among the most secure fasteners that you can use on a wooden run.

## Substrate

Once your birds are safely installed in your run, you'll notice that they quickly eat and scratch up every blade of grass or green plant in sight. Chickens can defoliate ground at remarkable speeds. Bare ground is just waiting to erode or to become mud, so you'll want to think ahead about what kind of substrate you'll use in your run when the time comes. We use untreated wood chips. They break down over time but do a good job in all sorts of weather, and the chickens like scratching them about.

For temporary fixes during snow and mud seasons, we have also resorted to wood shavings/bedding, piles of dried leaves, and even putting down pallets with rubber mats on top in some places. Some people swear by sand in their runs, while others use hemp or pine pellets, just like for horse stalls. See what is available in your area. As long as you don't use anything chemically treated or something that the chickens will eat and get impacted crops, you can experiment with substrate options.

*A downed tree led to a windfall of untreated wood chips to use as run substrate.*

## Soil Testing

If you are putting a coop and run on property with an old house or land that was used commercially, you will want to look into soil testing. Chickens, ducks, and geese are all very sensitive to heavy metals. They can develop a variety of serious symptoms and can die from heavy metal poisoning. If there was any chance that lead paint, leaded fuel, or overuse of commercial fertilizer contaminated your property, you should test your soil.

There are several universities throughout the United States that offer soil-testing kits. They are generally under $20 and are fairly quick and easy to do. The lab will send you the kits and instructions. Once your soil bags are filled, you mail them back to the lab and wait for results. Once you have your numbers, you'll be able to effectively modify your soil if necessary. Both lead and zinc poisoning can be treated, but it is far better to avoid the problem altogether and save yourself and your birds some heartache and pain.

## Alternative Run Spaces

If you will be keeping your flock in any kind of alternative space, such as a rooftop or a community garden, you will want to keep your birds safe from both predators and people. You want a very strong run and coop, and you may want to put locks on both of them. If the site has Wi-Fi, you could even place cameras in the run and the coop to keep an eye on things when you are not there.

## Predator Proofing

Everything wants to eat chickens. Foxes, coyotes, domestic dogs, bobcats, raccoons, weasels, owls, hawks, eagles, and even rats are all happy to have your chickens for dinner.

Your first line of defense against predators is to put a roof or top on your run. Even if you have 6- to 7-foot-high walls, aerial predators can still get in, and coyotes can jump astounding heights. Wooden runs can be built with wire or metal roofs, with support beams if necessary. T-post runs can be topped with lightweight wire or heavy-duty deer netting, with support poles to hold everything in place. Inspect your tops periodically to look for gaps and weak spots. If you have children, enlist their help with these inspections. It is important work, children love to be involved, and their short stature and curiosity often helps them spot problems before adults do.

An aviary-style run with a freestanding coop is one of the more predator-proof setups that you can have. Places where coops adjoin runs are weak spots that should be reinforced if at all possible, and checked often for security. Gates, coop doors, and any other openings are also potential weak spots that should be checked regularly. Make sure hinges are strong, and that doors and gates fit correctly in their framing and hang level. Try to minimize as much open space as possible around your gate.

Predators can fit under and through remarkably small spaces, and this includes gaps at the bottom of your run. To prevent anyone digging or squeezing under your wire fencing, it is a good idea to fashion an apron. Essentially, an apron is wire that is laid flat on the ground, forming a right angle to your run walls. Some people choose to dig down a couple of feet all the way around the run, bend the run wire to a right angle, and then bury it. You can also just lay wire on the ground around the perimeter of the run and build over the top of it. Over time, this wire gets covered and works the same as the buried wire. Even aboveground, the horizontal wire will help. The

third choice is to use blocks around the perimeter and/or the interior of the run to stop predators. Some of the steel-frame kits that we mentioned earlier have a gap between the frame and the ground. For these, cement blocks lining the bottom of the run inside and out is a good solution.

## Free-Ranging

Some of you may have been reading this chapter and thinking that you don't need a run—you're going to free-range your chickens every day. They'll be so happy and healthy! And, in most areas of the United States, they will be eaten by a predator before too long. Not if, but when.

Chickens do need to have time out of their run on fresh ground, but you have to be smart and careful about it. If your girls and boys are your beloved pets, we recommend building the biggest run possible, and then taking your flock out for an hour or so of supervised free-range as often as you can. Try to let your birds forage in spots that have plenty of coverage—bushes and shrubs, anywhere that they can dive out of sight—and away from any areas that known predators tend to frequent. You can even make a sort of permaculture habitat with secure fencing, lots of things planted for them to eat and hide in, and freedom for you to do some things without your eyes constantly glued to your flock.

Some chicken keepers believe that all chickens should be free-ranging all of the time, but if you choose to do so, you must be prepared for loss. It is our responsibility to care for our birds and keep them safe, and it can be devastating to lose birds that you love to predator attacks. In the end, you as the chicken keeper must decide the best way to balance safety with letting your chickens be chickens.

## Chicken-Safe Gardens

We could easily write another entire book on the topic of chickens and gardens. The image of gorgeous feathery creatures wandering near your flower beds and delicately scratching in the grass for grubs is very attractive. Unfortunately, the reality is much different. You can have chickens in your garden, and they will happily eat bugs, but you need to create barriers between your chickens and everything that you don't want them to eat, scratch up, or pull out of the ground. Raised beds help, especially the

Dandelions and marigolds are two of the most low-maintenance but rewarding plants that you can grow near your coops and runs. You can cultivate wild dandelions or plant seeds from named cultivars. Marigolds are very easy to grow from seed, and their pungent leaves act as a bug repellent. Chickens can eat both plants and will benefit from xanthophyll, yellow-pigmented carotenoids such as lutein and zeaxanthin, which help to give egg yolks a deep golden-orange color. Both plants also possess antioxidant, anti-inflammatory, and immune-supporting properties, as well as being a rich source of vitamins and minerals.

ones that are about 3 feet tall. You may have the odd light-bodied hen fly up into one, but in general, these beds are chicken-safe. (Safe from chickens, we mean.)

A couple of things that we've found helpful over the years if we want to release the chickens: covering vulnerable plants with wire "hats"; temporary fencing; and planting herbs or edible plants and flowers around the edges of the garden, or in big patches near the coops and runs. A lot of ornamental flowers are toxic to chickens, so double-check the list before you give your birds full access to your flower beds. Good options for chicken plantings include nasturtiums, marigolds, calendula, mint, oregano, sage, thyme, cress, and other greens.

Even though we don't let our chickens into most of our garden space, we do like to share our produce with them. Greens and herbs are some of their favorite things, and we'll toss them right into the run. You can grow shoots and sprouts in pots for your dear little dinos, and pop them into the run when they're ready to be devoured. If you have enough space, you can also grow melons or squash and pumpkins and share them with your flock, too.

## Poisonous Plants

***All Parts of Plant Are Toxic***

Angel's Trumpet

Avocado

Azalea and Rhododendron

Beans

Beardtongue

Boxwood

Castor Bean

Daffodil

Foxglove

Hydrangea—toxic in large quantities

Ivy—toxic in large quantities

Lantana

Lily of the Valley

Milkweed

Mountain Laurel

Nightshades

St. John's Wort

Sweet Peas

Wisteria

Yew

***Some Parts of Plants Are Toxic***

Cotton—seeds toxic in large quantities

Honeysuckle—berries are toxic

Poke Weed—berries and seeds are toxic, older leaves are dangerously high in oxalic acid

Rhubarb—leaves are extremely toxic

CHAPTER 6

# Our Favorite Heritage Breeds

We have arrived at our favorite part of this book: chicken breeds! We're passionate about chickens in general, but we love the variety of personalities, fancy feathers, and egg colors that you can get from heritage-breed chickens. We also love the history, and we'll share origin stories with you for the twelve breeds in this chapter. You'll also find practical information, like egg color, laying stats, appearance, and personality. You can source heritage breeds from hatcheries, private breeders, and farm supply stores. For more information about where we get our poultry, please see our list of resources at the back of the book.

If you choose to purchase your chicks from a farm supply store, you'll probably encounter some hybrid laying hens. As we mentioned in chapter 4, there are definite pros and cons to keeping hybrid layers. Just to recap: They don't breed true, so if you breed them you'll have barnyard mixes instead of purebred chickens. And that may be just fine for you. The hybrids tend to have fanciful names, like Green Queen and Prairie Bluebell Egger, and they're generally calm and docile. Because of their genetics, you will have a lot of eggs, but the trade-off is a shorter life span and a higher risk of reproductive diseases. For these reasons, we choose to concentrate on heritage breeds, and we suggest that our dear readers start their journey with a few well-chosen heritage-breed chickens.

In chapter 4 we covered the definition of heritage-breed chickens. They are breeds that have traditionally been kept in a certain geographical area over generations. They can be used in a breeding program to hatch purebred chicks that share the breed's characteristics and carry important genetics. The Livestock Conservancy is an American nonprofit organization that works to educate and bring awareness to

*Apple Blossom, a Delaware hen.*

the importance of heritage-breed livestock. They maintain a conservation priority list that puts the various animal species in conservation categories, from "recovering" to "critically endangered." For chickens, this list is based on poultry census data recorded by the Livestock Conservancy. Chickens that are critically endangered can move up or down the list, sometimes within a few years. In our breed descriptions we've specifically mentioned both breeds that have graduated from the list and breeds that are currently ranked as critically endangered. For the most current information, you'll want to visit the Livestock Conservancy's website.

Unlike hybrid laying hens, heritage-breed chickens generally live eight to ten years, and sometimes longer. One of the many things that we love about these birds is the fact that there's something for everyone. You can find a breed to fit just about any situation or inclination. And the best thing? The breeds we're about to describe to you are perfect for beginners, but they're just as loved by longtime keepers. It was hard for us to choose only twelve breeds. There are plenty of others around that would be great for beginners—Delawares, Jersey Giants, Langshans, Plymouth Rocks, and Dominiques, just to name a few. And a final word of warning: People tend to fall in love with their first breeds and keep them in their flock as long as they keep chickens. Choose wisely!

The American Poultry Association, or APA, is the governing body for sanctioned American poultry shows and stewards of accepted breed standards, including chickens, ducks, geese, and turkeys. Founded in 1874, the APA publishes the *Standard of Perfection* (SOP) for all accepted breeds. In our quick facts at the start of each breed section, we mention if a breed and any particular color variety or pattern of that breed has been accepted and included in the SOP. If you plan to breed your chickens, we strongly suggest that you get yourself a copy; it is invaluable when it comes to knowing how to evaluate show birds and selecting breeding stock.

## Orpington

### Quick Facts

**Country of Origin:** England

**Year(s) of Development:** 1880s

**Arrival in the United States:** 1891

**American Poultry Association Classification:** English

**Egg Color:** Light brown

**Egg Size:** Large/jumbo

**Average Egg Count:** 2–3 per week

**Personality:** Gentle, moderately active

*Buttercup, a Buff Orpington hen.*

**Outstanding Physical Features:** Very large chickens with lots of soft feathering

**APA Accepted:** 1902 for Buff Orpington, 1905 for the Black and White, and 1923 for the Blue

The Orpington is one of the best-known breeds among chicken keepers, both old and new. They're a heritage-breed success story and a very good option for brand-new Chicken Ladies and Gents. Orpingtons spent some time on the Livestock Conservancy's Conservation Priority List, and they graduated completely in 2016. Among other things, this means they are much more available to the public than they used to be.

Orpingtons are a heavy breed of chicken originating in Kent, England, and first introduced in 1886 in the black color variety. They are dual-purpose chickens (supply both eggs and meat), though not at all top-notch layers. We would call them respectable layers of brown eggs, but not good or great. Maybe we should call them triple-purpose, because they have always been very popular as show birds, too. In fact, we've read in many sources that the early Orpingtons were good layers, but generations of breeding for the show ring have bred out much of their laying prowess.

## Appearance

Orpingtons are big chickens. Roosters weigh in at around 10 pounds, while hens are around 8 pounds. They are large birds, and can look even larger with their beautiful, soft feathering. The feathers make their legs appear short, but they should never look as low to the ground as a Cochin, and their legs should always be free of feathers. They have a deep chest, and their back is a bit long and wide at the end. They have small heads in relation to their bodies, moderate-sized straight combs, bright red faces, combs, wattles, and earlobes, and white legs and feet. APA-accepted colors are Buff, Black, White, and Blue. Non-accepted but equally beautiful colors include Chocolate, Lavender, Red, Diamond Jubilee, Cuckoo, Lemon Cuckoo, Spangled, Silver Laced, Gold Laced, Partridge, and the exquisite Isabella pattern.

## Laying and Brooding

Young Orpington hens may produce 3–4 eggs per week, about 180–200 per year, but as they get older, this generally comes down to an average of 2 per week. These hens go broody often, and they are exemplary mothers.

## History

The Orpington breed was developed by William Cook of Orpington, Kent, England. Cook combined Black Langshan, Black Minorca, and Plymouth Rock chickens to create the Black Orpington. He quickly followed with the Buff Orpington, using Buff Cochins, Golden Spangled Hamburgs, and Dark Dorkings as foundation breeds. The beautiful White Orpingtons are supposed to be a sport, or natural color mutation, of the Buff.

William Cook was both a breeder and a businessman. He bred chickens, ducks (the Buff duck was one of his developments), and horses, and spent a lot of time and effort marketing them. He showed and marketed them in England, created the first Orpington Club, and eventually presented his birds at a meet at the Crystal Palace in London. But he didn't stop with breeding and showing. He developed and sold a line of poultry and horse medications, and even created a bird hospital. He sold feed, vitamins, delousing powder, broody/laying coops, poultry housing, and what sounds like everything you would need to keep chickens!

*Josie, a Buff Orpington hen.*

Almost the entire Cook family was involved with the poultry. Most of his children had a hand in the business, including a few of them who went on to establish international branches. Elizabeth Jane, the eldest daughter, ran the farm whenever her father was away from home, and she continued to manage the business for the rest of her life. Son Percy Cook set up and ran a branch of Cook & Sons in New Jersey. Another son ran a branch in South Africa. At some point, Cook packed up and took some of his ducks and his Black Orpingtons to sell in Australia. It was this line of Black Orpingtons that became the foundation for the Australorp—the Australian Utility Orpington.

Orpingtons had arrived in the United States by 1891, and they quickly started showing up at American Poultry shows. In 1903, William Cook packed up more of his stock and brought them to the United States. Americans appreciated the Orpington's excellent qualities. The Buff and the White were especially well received and added to Cook's success. Unfortunately, Cook's wife, Jane, died while he was abroad. He returned to England in 1904 and died shortly thereafter. His legacy is the creation and continuation of one of the most popular chickens ever known.

## What We Love About Them

Orpingtons are one of our favorite breeds. They are the quintessential English chicken and look perfect ambling around a garden. They are generally calm and excellent birds for families. This breed is friendly and likes being around people, especially if they are handled a lot while young. Our Orpington hens love being petted, and they're happy to snuggle down for a nap in a comfy lap. Orpingtons also do well in mixed-breed flocks.

If you want a broody hen to hatch and raise chicks for you, the Orpington is a first-rate choice. They're attentive mothers and take very good care of their babies. They're a great homestead breed that are as practical as they are beautiful.

Because of their large body mass, Orpingtons are not heat-hardy birds. Fans, shallow tubs of cool water to stand in, and plenty of deep shade will help them in hot weather. But all of the reasons why the heat bothers them are the reasons they do very well in the cold. They have lots and lots of soft feathers and down fluff that keep them warm.

# Cochin

## Quick Facts

**Country of Origin:** China

**Year(s) of Development:** China, Unknown; United States and UK, 1800s

**Arrival in the United States:** 1850s

**American Poultry Association Classification:** Asiatic

**Egg Color:** Light to medium brown

**Egg Size:** Large

**Average Egg Count:** 3 per week

**Personality:** Gentle, moderately active

**Outstanding Physical Features:** Large chickens with feathered feet and legs

**APA Accepted:** 1874, Buff Cochin

*Francesca, a Silver Laced Cochin hen.*

Cochins are the original fancy chicken. They originated in China before the 1800s and were further developed in the United Kingdom and United States. Chicken breeders and fanciers eventually produced the big, fluffy, gorgeous, feather-legged birds that are the modern Cochin. They're foundation breeds for so many chicken breeds all over the world that it would be hard to list all of them. Plymouth Rocks, Welsummers, Wyandottes, Orpingtons, Buckeyes, and Barnevelders are just a few.

Cochins are large, extremely well-feathered chickens that take a long time to grow and mature. They are used by some as a dual-purpose chicken, but the majority of those who keep and breed Cochins do so for showing and egg-laying pets. They come in a variety of gorgeous colors and are known the world over for their gentle personality and calm disposition. The Livestock Conservancy currently lists the Cochin in the Recovering category of the Conservation Priority List.

## Appearance

Cochins are large-sized fowl. Roosters weigh about 11 pounds, and hens weigh around 8.5 pounds. (They are supposed to be this big. It does not make them fat, so please do not withhold food from Cochins or any other heavyweight breed.) They have red faces, red earlobes, and moderately sized straight combs. Their legs and feet are yellow (in most color varieties, though a few have slate legs), but they're so heavily feathered that you usually don't see much more than the skin on their two inner toes. Their tails—both hens and roos—are short and wide, adding to their large, fluffy appearance.

Cochins come in a wide variety of beautiful colors and patterns. The APA notes in the SOP that the original Cochins were Buff in color, but the Buff was quickly followed by White, Black, and Partridge. Other Cochin color varieties include Silver Laced, Gold Laced, Blue, Brown, and Barred (or Cuckoo).

## Laying and Brooding

Cochin hens are respectable layers of large brown eggs. They lay around 180 eggs per year. Like most breeds, their numbers slow as they get older. They are top-notch broody hens and historically have been used to hatch both chicks and ducklings.

## History

Cochins were originally known as Shanghai Fowl because they were exported from the port of Shanghai to England and the United States. We found no clear answer about when and how the Cochins first arrived in the United States, though it was probably in the 1840s. Shanghais were documented as being in Boston for the first poultry show in 1849, and they appeared in the first printing of the APA SOP in 1874.

Queen Victoria is probably the most famous Cochin lover in history. She absolutely loved animals and kept a variety of poultry and fowl. Not long after she took the throne in 1837, Victoria commissioned a large aviary on the grounds of Windsor Castle. Sources say she spent a lot of time there with her different birds. However, it was the gift of five Cochin-China hens and two roosters by Edward Belcher, an officer in Her Majesty's Navy, that delighted her most.

In an interesting twist, the chickens that we call Cochins today were not the same birds known as Cochins in China. The famous Cochin-China fowl that first captivated

Queen Victoria were large birds much closer in appearance to a Malay, and may have had a completely different name in China, or wherever in Asia the chickens came from. The birds that we know as Cochins first showed up in the United Kingdom in 1847, on board a tea clipper ship docked in London. They were large, fluffy, and feather-legged. Though they were called Shanghai Fowl, for some reason people persisted in calling them Shanghai-Cochins, and later, just Cochins. It was this new "Cochin" that ignited Hen Fever in the United Kingdom and the United States.

*Francesca and Peony, Silver Laced Cochin hens.*

## What We Love About Them

These days, Cochins are generally still kept as pets, as show birds, and as good broody hens, especially for ducks and geese. They're big enough to fit those bigger eggs under them. They are so gentle and calm that they make excellent family pets, and do very well in a mixed flock of other gentle breeds. Because they're so fancy and not particularly predator-savvy, Cochins are probably not the best practical breed for a homestead, but they more than make up for it in looks and sweetness. They are plus-size supermodels and some of the best-ever hens to hug!

It wasn't just Cochins that were swept up in Hen Fever. Brahmas, Sebrights, Dorkings, and even White Faced Black Spanish were among the breeds selling for exorbitant sums of money. The upper and middle classes could not get enough of breed-specific chickens! The craze was fueled by the nineteenth-century interest in naturalism; by the opening of Asia to Western trade; and by the increase in wealth due to the Industrial Revolution, which meant more people could spend money on pet chickens.

## Brahma

### Quick Facts

*Eclair, Light Brahma hen.*

**Country of Origin:** China

**Year(s) of Development:** China, Unknown; United States and UK, 1850s

**Arrival in the United States:** 1850s

**American Poultry Association Classification:** Asiatic

**Egg Color:** Light to medium brown

**Egg Size:** Large

**Average Egg Count:** 3 per week

**Personality:** Gentle, moderately active

**Outstanding Physical Features:** Large chickens with feathered feet and legs

**APA Accepted:** 1874, Light and Dark Brahmas

The beautiful Brahma is one of the gentle giants of the chicken world. They have sweet and silly personalities and they make wonderful pets. They have well-feathered legs and profuse amounts of fluff. Brahmas were one of the breeds that ignited Hen Fever in Victorian England and America, selling for very large sums of money, and they remain a beloved breed with both the show community and backyard chicken keepers. In 2023, Brahmas graduated from the Livestock Conservancy's Conservation Priority List.

### Appearance

Brahmas are heavyweight chickens and are considered dual-purpose. The standard calls for roosters to weigh about 12 pounds and hens to weigh about 9.5 pounds. They have a small pea comb, red faces and red earlobes, and stout yellow legs. Legs and toes are heavily feathered, and tails are short and wide, like Cochins. They come in three color varieties: the Light Brahma, which is the Columbia pattern, a white bird with black striping on their necks and tails; the Buff, which is the Columbia pattern on buff feathers; and the Dark Brahma, a beautiful steel-gray and silver penciled bird with a light head and neck. Photos do not do them justice. They are stunning birds.

## Laying and Brooding

Brahma hens are very good broody mamas! They're decent layers of large brown eggs. They generally do 3–4 per week their first few years of laying, then drop off to 2–3 per week as they get older. They're good winter layers and will start laying again in January or February, after their annual molt.

## History

This bird has a convoluted history. Brahmas, or the original stock that was developed into what we know as Brahmas, arrived in the United States and the United Kingdom about the same time, in 1850. These various early versions of Brahmas were called Chittagongs, Gray Chittagongs, and Brahmaputras. Some of this stock did come from a ship that had been docked in the Brahmaputra River, and the chickens were given the same name. But there are also sources that claim some of the stock came from Shanghai, and may have included birds that were the result of a cross between Cochins (Shanghai Fowl) and Malays, or birds very similar to the Malay. Either way, by 1874, they had a breed standard and a shorter name, and both Light and Dark Brahmas appeared in the APA's first printing of the SOP in 1874. The beautiful Buff Brahmas were developed and added to the SOP later, in 1924.

Queen Victoria did have a pair of Brahmas along with her Cochin-Chinas and Shanghai Fowl. An American breeder, George Barnham, presented Her Majesty with a pair, or maybe a small flock, of Light Brahmas, and used all of his marketing skill to promote this gift, thereby allowing him to raise the price of his Brahmaputra birds, until they were selling for as much as $150 a pair. Light Brahmas are believed to be the foundation stock for the Dark Brahmas, and both varieties were extremely popular in the United States and United Kingdom. In *The New American Poultry Book*, published in 1888, John Taggart describes Brahmas as being the largest breed of chicken; it was probably true then, before the Jersey Giant had been developed.

*Crêpes Suzette, a Buff Brahma hen.*

## What We Love About Them

Brahmas are fancy chickens and make excellent additions to the Chicken Lady or Gent's flock. They also make great family birds and wonderful members of an easy-going mixed flock. They're not good candidates for rough-and-tumble homesteads. Unsupervised free-range is dangerous, especially for the Light Brahma; she makes a large, bright white target. They like to forage and scratch, but they also like long, lazy dust baths and afternoon naps. They are excellent lap chickens, and wonderful to hug!

Brahmas are very cold-hardy, but do not enjoy summer. They need lots of shade, cool water, fans—especially in the coop at night—and even shallow pans of cool water to stand in.

## Australorp

### Quick Facts

**Country of Origin:** Australia

**Year(s) of Development:** 1890s/1900s

**Arrival in the United States:** 1920s

**American Poultry Association Classification:** English

**Egg Color:** Brown

**Egg Size:** Large

**Average Egg Count:** 4–5 per week

**Personality:** Gentle and calm

**Outstanding Physical Features:** Glossy black with green highlights!

**APA Accepted:** 1929

*Hideo and Katsumoto, Australorp roosters.*

The gorgeous Australorp—or Australian Black Utility Orpington—was created in Australia in the late 1800s and early 1900s, principally for egg production, but also as a dual-purpose heavy breed. The Livestock Conservancy currently lists the Australorp in the Recovering category.

## Appearance

Australorps are large-bodied chickens, glossy black with a beetle-green sheen. They have moderately large straight combs and wattles and red earlobes. Their elegant appearance is helped along by dark brown eyes, a dark beak, and dark legs with pink soles on their feet. Roos weigh around 8.5 pounds, and hens, 6.5 pounds. There is a bantam version.

## Laying and Brooding

Australorp hens are excellent layers of large brown eggs. Despite the fact that they are such good layers, Australorps go broody and make good mothers. Broodiness will cut down on egg numbers, but you won't necessarily need an incubator if you want to hatch eggs.

## History

There are several vague accounts of the origins of the Australorp, so we've done our best to weed through and put together the story for you from the most reliable accounts and sources. The original Orpington chickens developed by William Cook were black. Some of these black Orpingtons were imported into Australia from around 1890 to 1900. The Livestock Conservancy notes that about this time, the English breeders were trying to refine the Orpington into a meat bird. The Australians, though, looking for both food security and business opportunities, wanted to concentrate on egg production. To do this, they crossed the Black Orpington with some of the foundation breeds that went into its early development: the Minorca, White Leghorn, and Langshan. One of the goals of the Australians was to breed out the Orpington's broody tendencies to produce higher egg numbers.

Here's where things get interesting. From 1902 to about 1924, the Australorp took part in several egg-laying contests that were held in Australia. The first contest was held by an Australian agricultural college, and Orpington hens took seven places out of the top thirteen. In another contest, a team of sexed Australorp hens laid 1,857 eggs over a one-year period. That's just over 309 eggs each. The Livestock Conservancy notes that the Australorp holds the world record for laying eggs: 364 eggs in 365 days!

The Australorp was a great success as a good-sized, dual-purpose chicken that truly was a champion layer. News of the record-setting hens traveled to the United Kingdom and United States, and Australorps showed up in both countries during the 1920s.

### What We Love About Them

The Australorp is a fantastic homestead bird, great with kids, and a wonderful addition to a mellow layer flock. They're gentle and friendly birds, beautiful in an elegantly simple way. They're pretty cold-hardy, but can also handle some hot weather. All the same, in the heat they do need shade, cool water, and fans. Roosters have big combs and wattles that need to be protected from frostbite in frigid weather (see chapter 9 for more information on frostbite prevention). Because of their overall awesomeness, we feel that the Australorp should be better known and more widely kept.

## Salmon Faverolles

### Quick Facts

**Country of Origin:** France

**Year(s) of Development:** 1850s–1890s

**Arrival in the United States:** Early 1900s

**American Poultry Association Classification:** Continental

**Egg Color:** Light brown

**Egg Size:** Small–medium

**Average Egg Count:** 2–3 per week

**Personality:** Very gentle

**Outstanding Physical Features:** Beard, muff, feathered legs and feet

**APA Accepted:** 1914, Salmon; and 1981, White

*Angelica, Salmon Faverolles hen.*

The ridiculously adorable Faverolles chicken was developed in France. They're a heavily feathered, medium-sized chicken that was developed to be a true dual-purpose breed. Fifteen years ago Faverolles were critically endangered, but they're currently in the Watch category of the Livestock Conservancy's Conservation Priority List. These little chickens are much too special to go extinct!

## Appearance

Faverolles are a medium-bodied chicken with feathered legs, full muff (feathers that extend up from the beard and give the appearance of earmuffs or very fluffy cheeks) and beard, a straight comb, white skin and legs, and five toes. In the Salmon variety, hens are a beautiful mix of salmon, brown, and cream shades. Roosters have a black beard and their front is black as well. They look very different from the hens but are just as cute. The APA SOP puts their ideal weight at 8 pounds for roosters and 5.5 pounds for hens.

## Laying and Brooding

Faverolles hens are pretty good layers, averaging about 180 medium to large light brown eggs per year. Faverolles hens will occasionally go broody. Reports say they're relatively easy to break from broodiness, however. Faverolles are not a typical heritage breed in that they mature early and will begin to lay earlier than a lot of other heritage breeds.

## History

Faverolles were developed in France around the village of Faverolles, an area south of Paris. They were intended to be a true dual-purpose breed. Farmers were trying to create a chicken that was an excellent table bird plus a good winter layer, so they could have more products for sale at the Paris winter market. The British Faverolles Society notes that many of the French farmers were also trying to develop a breed that would tolerate confinement or caging well.

French farmers and breeders spent most of the second half of the nineteenth century working on the Faverolles. Among the foundation breeds were Houdans,

Brahmas, and Dorkings. The Livestock Conservancy notes that French Rennes, Flemish Cuckoos, and Malines may also have been included. The Faverolles' fifth toe was almost certainly a gift from the Dorkings. The Houdan supplied the beard and muff, and the Brahma passed along the feathered legs and feet. In the Salmon variety, the males' duck-wing color pattern is another gift from the Dorking, while the hens' salmon color is unique to chickens and only occurs in this breed. It took a few decades of work before the Faverolles started to have a uniform appearance. At some point during the 1880s, Faverolles began to look like the birds that we know and love.

In the 1890s the Faverolles arrived in the United Kingdom and quickly became a favorite breed. Faverolles arrived in the United States in the very early 1900s, imported by Dr. Phelps of New York. Phelps placed an ad in a 1904 poultry journal advertising his Faverolles, imported directly from France and England by himself, which was a strong selling point. Dr. Phelps also worked with a few other breeds and occasionally wrote articles that were published in poultry journals of the time.

## What We Love About Them

Faverolles are sweet and docile, but they're also active, extremely alert, and have lots and lots to say! We love them for their adorable appearance and their gentle ways. Be wary of putting them into an already-established flock with strong personalities. This breed is very cold-hardy, but they do not like the heat and need help to keep cool. One of the dangers of having them in a flock full of strong personalities is that they could be chased or picked on in hot weather.

Faverolles make very good show chickens and have been found in show circles since the late 1800s. These adorable chickens are not at all predator-savvy, and may even need their muffs trimmed away from their eyes occasionally, to prevent obstructed vision. While they're healthy and hardy chickens, they're not super rough-and-tumble, so reconsider if you want a hands-off homestead breed. They are fantastic in a relaxed flock of other gentle breeds. We think every Chicken Lady or Gent should have a couple of Faverolles in their flock for added style and the softest chicken hugs.

## Welsummer

### Quick Facts

**Country of Origin:** Netherlands

**Year(s) of Development:** Late 1800s

**Arrival in the United States:** Mid-twentieth century

**American Poultry Association Classification:** Continental

**Egg Color:** Terra cotta brown, sometimes speckled

**Egg Size:** Large

**Average Egg Count:** 3–4 per week

**Personality:** Smart, curious, active, and friendly

**Outstanding Physical Features:** Beautiful red partridge feathers

**APA Accepted:** 1991

*Gertie, a very special Welsummer hen.*

The Welsummer is a Dutch breed developed in the village of Welsum in the Netherlands. The Welsummer's primary purpose was (and still is) their richly colored brown eggs. Beloved birds in their native country, a pair of Welsummer statues stand in front of the church in Welsum. King Charles used to keep a breeding flock of them at his home in the United Kingdom. This European heritage breed has also become quite popular in the United States.

### Appearance

The Welsummer is considered a large fowl here in the United States. Roosters should average around 7 pounds, hens, around 6 pounds. Their beautiful coloring is known as red partridge. It is a colorful pattern, with plenty of reddish-brown feathers laced with black, and red-gold neck feathers also laced with black. Welsummer roosters are even more colorful. They have moderate-sized straight combs, red earlobes, red faces, clean yellow legs, somewhat tight feathering, and intense reddish eyes. There is a bantam version, but their eggs aren't as dark as those of the large fowl.

## Laying and Brooding

Welsummer hens are very good layers of reddish-brown, and sometimes speckled, eggs. They can produce around 180–200 eggs per year. They are known to be non-setters and generally will not go broody.

## History

Welsummer development started around the village of Welsum in the late 1800s. The origins were probably a particular strain of country, landrace, or farm bird that someone started breeding for "improvements," such as more eggs or a larger body. Some of the farmers around Welsum wanted a brown egg layer—either for profit, or because they wanted something different than the white eggs that so many of the Dutch breeds lay.

In the early decades, there was a lot of disparity in the appearance of the Welsummer. To establish a breed standard and help support the farmers working with this chicken, a breed association was formed in the Netherlands in 1928. This helped to achieve a more consistent appearance as well as supporting their superior laying ability. The Welsummer only comes in one color variety here in the United States: Red Partridge. It will come as no surprise that foundation breeds include Partridge Cochin, Partridge Wyandotte, Partridge Leghorn, Barnevelder, and Rhode Island Red.

Welsummers arrived in the United States sometime in the early to mid-twentieth century, but did not gain much popularity until they were APA-accepted. Many of the bloodlines here in the United States were imported as hatching eggs, but there was also importation of English bloodlines through Canada.

## What We Love About Them

One of our favorite things about the Welsummer is their excellent personality. They are friendly, very talkative, and great with kids and families. Because they get along well with both gentle hens and those with a strong personality, they're right at home in a mixed flock. They make a great homestead breed and lay pretty eggs. These are intelligent birds, very good foragers, excellent show chickens, great in the garden, and very, very good at being beloved pets. They are also quite hardy and healthy birds. Welsummers do well in most climates, though their combs and wattles may require protection from frostbite in the winter (see chapter 9 for more information). Make sure they have opportunities to free-range so they don't get bored. This chicken is the perfect combination of beautiful and practical.

## Barnevelder

### Quick Facts

**Country of Origin:** Belgium

**Year(s) of Development:** Late 1800s

**Arrival in the United States:** Mid- to late twentieth century

**American Poultry Association Classification:** Continental

**Egg Color:** Dark brown

**Egg Size:** Large

**Average Egg Count:** 3–4 per week

**Personality:** Gentle and friendly

**Outstanding Physical Features:** Glossy brown and black laced feathers

**APA Accepted:** 1991

*Truffle, a Barnevelder hen.*

We believe that the Barnevelder is one of the most perfect chickens in existence. They are a large, dual-purpose, European heritage breed. They are beautiful, have great personalities, and lay pretty eggs. Once hard to find, Barnevelders are becoming more popular in the United States. While perfect for beginning chicken keepers, they also have plenty to offer the more experienced as well.

## Appearance

Barnevelders are a standard-sized, dual-purpose breed. Roosters weigh around 7 pounds, hens, around 6 pounds. They have medium-sized straight combs, with shorter combs on hens, and moderately sized wattles. They have red faces with a lot of feather coverage, red earlobes, and clean yellow legs. There are two color varieties available in the United States: the gorgeous brown-and-black, double laced partridge, and the not-yet-APA-accepted beautiful silver double laced, imported by Greenfire Farms in 2016.

## Laying and Brooding

Barnevelder hens are good layers of medium to dark brown, often-speckled eggs. They lay around 200 or so per year. They very occasionally go broody, but we would not consider them a broody breed.

## History

The Barnevelder is a fairly modern Dutch breed, developed in the Barneveld region of the Netherlands. In the late 1800s, Dutch farmers began breeding Asiatics—Cochin, Brahma, and Langshan—with their local country hens. They may have been trying to increase size for dual purpose, they may have been trying to create show birds, or they may have just been experimenting. It is also possible that, like the Welsummer, breeders were trying to develop the Barnevelder as a colored egg layer.

It took several generations before the Barnevelder started to emerge with a standard appearance. In the 1920s, the Dutch poultry club accepted the breed, and they were exported to the United Kingdom. We have been unable to find the exact date when Barnevelders arrived in the United States; our best guess is mid- to late 1900s.

## What We Love About Them

These birds are gorgeous, with brown-and-black laced feathers. They are both sweet and smart, and make great additions to a mixed flock. Barnevelders are good layers of pretty eggs and make excellent pets and companions. You can show Barnevelders, and because of their mellow personalities, they would be perfect birds for 4-H kids. We are in love with these wonderful birds.

Barnevelders are hardy and healthy chickens. They do well in the cold, though you should protect those combs from frostbite. They do okay in the heat, with plenty of shade and cool water and other provisions (like a nice, chilled watermelon!), but they don't love it. They are active birds and very much enjoy time free-ranging and foraging. It's somehow quite soothing to sit outside and listen to your chatty Barnevelders cluck to each other as they dig for bugs. That's Chicken Lady heaven.

## Crested Cream Legbar

### Quick Facts

*Beatrix Potter, a Crested Cream Legbar hen.*

**Country of Origin:** United Kingdom

**Year(s) of Development:** 1930s

**Arrival in the United States:** 2010

**American Poultry Association Classification:** None

**Egg Color:** Blue, green, turquoise

**Egg Size:** Medium–large

**Average Egg Count:** 4 per week

**Personality:** Active, curious, friendly

**Outstanding Physical Features:** Small backwards crest

**APA Accepted:** Not at this time

The pretty little Crested Cream Legbar is another one of our favorite breeds. They were developed in the United Kingdom in the 1930s at Cambridge University. They're small-bodied chickens with a small, backwards crest and big personalities. Legbars are auto-sexing—males and females can be identified by different colors as chicks—so they're an excellent choice for someone who cannot have roosters but still wants some beautiful turquoise eggs.

## Appearance

Legbars are light-bodied chickens with small, swept-back crests, straight combs—sometimes large and ruffly—small wattles, white earlobes, and yellow legs and feet. The hens are a mixture of beautiful earth shades: browns, grays, and creams. They sometimes have stippling or barring show up in their neck feathers and other parts of their plumage. The roosters are similar colors, but with a lot more barring on their bodies and tails. Though the Cream is by far the most popular color, there are also Golden Crele (a barred feather pattern in colors other than black and white) and White varieties in the United States.

## Laying and Brooding

Legbars are good layers of blue (occasionally green) eggs. Pullets start laying at around six months and will lay an average of 4 medium to large eggs per week. We've heard rumors that an occasional white egg layer will show up, but we do not know anyone who has had this happen. Legbars will go broody occasionally, but it is rare. They may become broody more often when kept around other breeds that are broody a lot.

## History

In 1927, Clarence Elliot, a British horticulturist, traveled to South America to collect "exotic" plants and animals. The United States, the United Kingdom, and several other countries funded expeditions of this sort, hoping that the collectors would bring back seeds or animals that would help with the many agricultural problems that cropped up after World War I and into the 1920s. Among the exotics Elliot brought home was a group of chickens that just happened to lay blue eggs. The chickens were given to Reginald C. Punnett, the director of the Cambridge University Breeding Program, essentially the leader in genetic research in the UK, and Punnett was one of the foremost genetic scientists in the world. He gave biology students everywhere the gift of the Punnett Square.

It seems funny that such an endearing little bird was created in an academic program, but that's where the Legbar came from. Punnett's work was focused on improving the livestock raised on British farms. He and his team were working on sex-linked genes and the creation of auto-sexing breeds. They crossed one of the Chilean hens—"a crested yellow hen"—with a golden-penciled Hamburg, and then

*Cornelia, a Crested Cream Legbar hen.*

crossed some of their offspring. A Brown Leghorn cockerel was eventually added to the breeding program, with the resulting chicks carrying a gene for cream coloring. One of Punnett's associates, Michael Pease, had crossed a White Leghorn cockerel with some of the program's Gold Legbars. The offspring also carried the identical cream gene. Punnett and Pease bred these lines together and continued to select for crests, auto-sexing, and blue egg laying. The Crested Cream Legbar was presented to the public at the 1947 London Dairy Show.

The British Poultry Club officially recognized the Cream Legbar in 1958, but twenty years later it had already reached Threatened status. Industrial egg and chicken farms had grown, and backyard chicken keeping was much less popular. Also, at the time there was no market for "novelty eggs." As we know, this changed about twenty-five years later when Ameraucanas and Easter Eggers became popular. Legbars are designated a rare breed in the United Kingdom by the Rare Breeds Survival Trust. They are not currently accepted by the APA.

Legbars have only been in the United States since the fall of 2010, imported by Greenfire Farms in Florida. Most of the first batch perished due to the US strain of Marek's disease, a deadly virus that attacks a chicken's immune system. The surviving trio was resistant to Marek's, and was used to build up the US population. They made another importation of unrelated stock in 2011, and had those birds vaccinated for Marek's. In 2013, they made a third importation of Legbars, this time to serve as stock for champion Legbar breeder Jill Rees.

## What We Love About Them

Descriptions of the Legbar personality vary: Some say they are nervous and flighty, while others say they're friendly and docile. You get out what you put in. Our Legbars are friendly and curious and love to free-range and forage. They tend to be toward the bottom of the pecking order, so watch your flock dynamics and avoid putting them in with a strong flock where they'll get picked on. Legbars make good homestead birds in many ways, as well as garden helpers, good layers, and being gentle with kids. Add it all together and the Legbar is a true Chicken Lady favorite!

Legbars are small-bodied birds and, like their ancestors, the Leghorn, not really cold-hardy. Their combs definitely need protection from frostbite, especially the lovely ruffled combs.

## Speckled Sussex

### Quick Facts

*Mary Berry, a Speckled Sussex hen.*

**Country of Origin:** United Kingdom

**Year(s) of Development:** Unknown; "improvement" in the late 1800s

**Arrival in the United States:** Possibly Colonial America; officially, 1912

**American Poultry Association Classification:** English

**Egg Color:** Light brown

**Egg Size:** Large/jumbo

**Average Egg Count:** 3–4 per week

**Personality:** Outgoing, friendly, smart

**Outstanding Physical Features:** Gorgeous glossy spots on the Speckled

**APA Accepted:** 1914, Speckled and Red; 1929, Light

If you're new to chickens and you could only ever pick one breed as your forever chicken, consider the Sussex. They're gorgeous and have great personalities, they're good foragers, and they're a dual-purpose heritage breed with a long history. Sussex chickens are currently listed in the Recovering category of the Livestock Conservancy's Conservation Priority List.

### Appearance

Sussex are classified as a heavy breed, with roosters weighing in at about 9 pounds, and hens, around 7 pounds. That's a nice armful of chicken! They have moderately sized combs and wattles, white legs and feet, and red faces and earlobes. They have long, wide backs and legs just slightly on the short side. Here in the United States they come in three color varieties: Speckled (white and green/black spangles on a deep maroon background), Red (beautiful deep red with a dark head), and Light (Columbia pattern). Non-APA-accepted colors include white, brown, silver, buff, and Coronation. (The Coronation Sussex was named for the coronation of King George VI of England, Queen Elizabeth II's father, in 1937. This color variety has blue/gray feathers in a modified Columbia pattern.)

## Laying and Brooding

Sussex hens are very good layers. They'll produce an average of 200-plus large brown eggs per year, sometimes speckled, sometimes pink in shade. They will sometimes go broody, and make very good mothers when they hatch a clutch.

## History

Sussex were developed in England in Sussex County and the surrounding region. They're considered a heavy breed, and many sources claim they've been around since the Romans invaded England. It is difficult to find much written about their early history. Some contemporary poultry historians suggest that the Sussex may have already been in England when the Romans brought in the Dorkings. The Sussex tend to be lumped in with the Dorkings by early poultry historians. Both breeds were developed in the same general area and have some similar characteristics. We've found many references to Speckled Dorkings, and believe they may have been Sussex.

British poultry historian Lewis Wright has a whole chapter on Dorkings/Sussex in *Wright's Book of Poultry*, published in 1885. Wright mentions the Sussex as a separate breed from the Dorkings, but one that was in great danger of going extinct. The Sussex was accepted by the British Poultry Club in 1902, and a Sussex Poultry Club was started in 1903. Wright, whose personal favorite was the Light Sussex, was an active member. The club was successful in bringing recognition to the Sussex and establishing them as an important breed on British farms and smallholdings. The Sussex is believed to have been present in Colonial and Early America, though there is no hard evidence. For this reason, their documented "official arrival" in the United States is 1912.

## What We Love About Them

Sussex chickens are reasonably climate-hardy. They manage fairly well in both warm and cold climates. Like the other breeds with moderate to large straight combs and wattles, they'll need frostbite protection. While they can tolerate some heat, they are big chickens, so shade and fans and cool water are important during hot spells. Sussex hens tend to be reliable layers despite weather conditions.

This is a great homestead breed. They are good foragers and do a great job turning over compost and beds in a winter garden or orchard. Sussex are also perfect for families, good with kids, super friendly, and excellent in a mixed flock. We like to say that they're smart, but not too smart. Our Sussex hens are outgoing and cuddly—an ideal chicken in many ways. They're often the first at the gate to say hello. We should note that they're rather chatty chickens, too.

## Rhode Island Red

### Quick Facts

**Country of Origin:** United States

**Year(s) of Development:** Late 1800s

**American Poultry Association Classification:** American

**Egg Color:** Light to medium brown

**Egg Size:** Large/jumbo

**Average Egg Count:** 4–5 per week

**Personality:** Confident, outgoing, friendly, active

**Outstanding Physical Features:** Beautiful red feathers

**APA Accepted:** 1904, Straight Comb; and 1905, Rose Comb

*Spicy, a Rhode Island Red hen.*

The Rhode Island Red (RIR) is the quintessential American chicken, and is arguably the most famous chicken in the world. A true dual-purpose breed, they were a mainstay of homesteads and small farms throughout most of the twentieth century. During the latter half of the twentieth century, a commercial strain of RIRs was developed. We are concentrating on the larger, darker-red heritage strain of the Rhode Island Red. The Livestock Conservancy currently lists non-industrial heritage lines of RIRs in the Watch category of the Conservation Priority List.

## Appearance

Heritage Rhode Island Reds come in two types: Straight Comb and Rose Comb. Both types are a lustrous, deep maroon red with black tail feathers. They have bright red faces and earlobes, bright yellow legs, and a strong, blocky body. Roosters average about 8.5 pounds, hens, about 6 pounds.

*Spicy!*

## Laying and Brooding

RIR hens are excellent layers of large brown eggs, often 250–300 per year. They will go broody occasionally.

## History

Here's what you usually read: RIRs were developed in New England during the late nineteenth and early twentieth centuries. The APA SOP included the Straight Comb RIR in 1904 and Rose Comb in 1905, and lists the Red Malay, the Brown Leghorn, and "Asiatic Native stock" as the foundation breeds. Other sources list the Shanghai chicken, which may be the Cochin or may be the Malay-like precursor to the Cochin. The Livestock Conservancy also lists the Java as a foundation breed, specifically the Auburn Java.

The story of the Rhode Island Red starts with a fishing captain. Sometime in the early 1850s, before the Civil War, Captain William Tripp acquired a Red Malay cockerel. He took it home and let the cockerel start breeding with his hens. The resulting offspring pleased the captain. They reportedly laid more eggs and worked very well as dual-purpose birds.

Captain Tripp presented his friend John Macomber with some of the Malay-cross chickens to take home and breed with Macomber's own chickens. These two friends decided to use these Malay-cross birds to start their own breeding projects. They occasionally traded birds back and forth to minimize inbreeding, and they did occasionally bring in other breeds: Brahmas, Javas, and Cochins. At this point in time the birds did not have a standard appearance and were producing offspring with a mix of straight, rose, and pea combs. Eventually, word got out that these two men had an improved breed of chicken, and local farmers began buying breeding stock from them. The birds were known as Macombers, or Tripp's Fowl.

One of the farmers who came to Tripp and Macomber to buy the improved stock was a man named Isaac Wilbour. Wilbour ran a large egg farm in Little Compton, Rhode Island, supplying eggs to the Boston and Providence, Rhode Island, markets. He acquired some of the Macomber/Tripp's Fowls for their reputed laying ability. Wilbour was the grandson of the first Isaac Wilbour, one of the early governors of Rhode Island. He was very well connected financially and socially, and the person who really put the Rhode Island Reds on the map.

Wilbour wanted to continue producing a dual-purpose bird, but he also wanted them to be great layers to support his egg business and large enough to deal with the cold New England winters. There is a theory that Wilbour crossed his Tripp's Fowl with Brown Leghorns—which arrived in the United States in 1853—for maximum egg production. He carried on with breeding, and in the late 1890s, two professors from the Rhode Island Agricultural Experiment Station, now known as the University of Rhode Island, came to visit his farm to see the breed. They were impressed with his birds and asked him to name them. He chose "Rhode Island Red."

The breed became super popular and quickly spread all over the United States. Egg farmers, smallholders, and backyard chicken keepers all had their flocks of Rhode Island Reds.

## What We Love About Them

Rhode Island Reds are bright, curious, alert, and intelligent birds who can be quite friendly if handled a lot when young. They can have strong personalities, especially in large groups, so plan carefully when mixing them with gentler breeds. This is an excellent breed for almost any purpose: egg laying, showing, homesteads, pets, compost/garden work, etc. Like most American breeds they are reasonably cold-hardy—watch combs and wattles for frostbite—and can withstand some heat. They are one of the best layers among the heritage breeds, second only to the Leghorn and Australorp.

# Ameraucana and Easter Egger

## Quick Facts

**Country of Origin:** United States

**Year(s) of Development:** 1930s–1970s

**American Poultry Association Classification:** American

**Egg Color:** Blue, green, light brown

**Egg Size:** Large

**Average Egg Count:** 3–4 per week

**Personality:** Friendly, active, sometimes skittish

**Outstanding Physical Features:** Full beard and muff

**APA Accepted:** 1984

*Violet, a young Ameraucana hen.*

Ameraucanas are one of the more popular breeds in America. They're super cute with their muffs and beards, they're gentle chickens, and most of them lay beautiful blue and green eggs. They are a distinct breed with a breed standard. We're emphasizing that Ameraucanas are a true breed to differentiate them from Easter Eggers, who are not a true breed. Ameraucanas and Easter Eggers are plentiful and not in need of conservation. Easter Egger is a general term for mixed-breed chickens that carry the

Easter Eggers were developed in the 1970s. Cackle Hatchery, who has been working with them since 1971, holds a trademark on the name "Easter Eggers." These birds are produced from a mix of breeds, some of whom must carry blue-egg-laying genes. There is no standard appearance for Easter Eggers. They do not have distinct DNA, nor do they breed true. But they are generally friendly, healthy, and long-lived birds who lay well and are good for beginning chicken keepers.

gene for laying blue/green eggs. They do not have a breed standard or a consistent appearance, but like many "mutts," they're often quite healthy and robust.

*Blanche DuBois, a green-egg-laying Easter Egger.*

## Appearance

Ameraucanas are adorable birds that come in a wide variety of colors. The breed standard calls for muffs and beards and tails. No muff and beard, no tail—no showing that bird. They're medium-bodied in size, have a pea comb, and virtually nonexistent wattles. Their legs and feet tend to be slate, black, or occasionally blue in color. Tails are usually upright at about 45 degrees. Roosters usually weigh 5.5–6.5 pounds, hens, 4.5–5.5 pounds. The roosters tend to be very handsome boys.

The APA recognizes ten gorgeous color varieties: Black, Blue, Blue Wheaten, Brown Red, Buff, Silver, Wheaten, White, the newly added and very pretty Self Blue (what we would generally call Lavender), and the equally pretty Splash. Some hatcheries and non-competition breeders tend to keep Ameraucanas in mixed breeding flocks—not separated by color. The result is generally some beautifully unique feather patterns and colors.

## Laying and Brooding

Ameraucana hens are pretty good layers of blue, green, or light brown eggs. They'll lay around 180–200 eggs per year. They generally will not go broody.

## History

Ameraucanas are relative newcomers to the standard bred poultry world. They've been around since the 1970s, largely derived from Araucana stock and some other unknown mix of heritage breeds. Araucanas, unusual-looking blue-egg-laying chickens with curly ear tufts and no tail, arrived in the United States from South America as early as 1930. At that point, most people working with the breed were trying to keep Araucana genes, like the blue eggs, but mix in enough other DNA to get rid of the lethal ear tuft genes and any other inherited problems. (Araucanas carry a gene for ear tufts that is tied to in-shell chick deaths when offspring carry a double copy of the gene.) Ameraucanas were the very successful result of those breeding projects. They are not considered an at-risk breed.

In the early days, breeders were working on both bantam and large fowl Ameraucanas. The bantams actually paved the way for acceptance of the standard-sized birds. In 1978, the Ameraucana Bantam Club was formed with the object of creating a breed standard and gaining admittance into the APA and American Bantam Association. In 1980, the ABA did accept Ameraucana into their SOP. While the club concentrated on bantams at first, there were breeders working on large-fowl Ameraucanas as well. In 1984, the APA accepted both the bantam and large-fowl Ameraucanas into the SOP. In 1986, the name of the Ameraucana Bantam Club was changed to Ameraucana Breeders Club, to reflect the members who were committed to the large-fowl Ameraucanas as well.

## What We Love About Them

Ameraucanas are very nice additions to a mixed laying flock, and beginning chicken keepers often get them. They're gentle and easy to deal with, though some are a bit nervous and even skittish at times. The muffs can restrict their vision a bit, so we recommend supervised free-range time and/or carefully trimming feathers away from eyes. These birds can be good garden helpers, and they usually get along with their flock mates pretty well. They're not usually high in the pecking order and may be bullied by stronger breeds, so keep an eye out for that. They're not generally used as a homestead breed, but they are considered dual-purpose by some. Most of the time they're pets and companions, beloved members of backyard flocks, or show birds. Their beards and muffs and tiny combs and wattles make them fairly cold-hardy and not prone to frostbite. They are relatively adaptable to both heat and cold—but watch them in either extreme.

## Bielefelder

### Quick Facts

**Country of Origin:** Germany

**Year(s) of Development:** 1970s

**Arrival in the United States:** 2011

**American Poultry Association Classification:** Not accepted

**Egg Color:** Brown

**Egg Size:** Large/jumbo

**Average Egg Count:** Around 4 per week

**Personality:** Gentle and friendly

**Outstanding Physical Features:** Large birds with beautiful golden barring

**APA Accepted:** Not at this time

*Little Jackie, a Bielefelder hen.*

Crele is a color pattern that features barring over the top of black-breasted red coloring. It can have red, chestnut, and/or gold accents.

The Bielefelder is a large, beautiful, Crele-patterned chicken that lays a lot of eggs and has a sweet and friendly personality. Sometimes called the Uber Chicken in their native Germany, the Bielefelder was developed in the 1970s as a dual-purpose auto-sexing breed. While they meet all of the criteria for a heritage breed, they have not yet been accepted by the APA. There is a German breed standard, and a Facebook breeders' group to help you breed to the German standard, should you want to do so.

### Appearance

Bielefelders are very large birds; roosters average about 9ish pounds, hens, about 7ish pounds. They have moderate to large straight combs, red earlobes, and yellow legs. Bielefelders are beautiful, colorful chickens. A second color has been developed—silver—which is also barred.

## Laying and Brooding

The very good-natured Bielefelder hens lay 220 or more large brown eggs per year. The eggs are often speckled, like Barnevelder eggs. These hens rarely go broody, but there are some reports of occasional broodiness. They also mature early and start laying sooner than many heritage breeds.

## History

These beautiful chickens were originally bred in the North Rhine / Westphalia region, in the town of Bielefeld, Germany, by a man named Gerd Roth. He crossed French Malines and Welsummers with Barred Rocks. The auto-sexing comes from the Barred Rocks, while the Crele pattern comes from the barred Malines and partridge-colored Welsummers. We've also seen the New Hampshire mentioned as a foundation breed, but so far have found no hard evidence.

Bielefelders were first shown in Germany in 1976, and at that time they were called Deutsche Kennhuhn—roughly translated to something romantic-sounding, like "auto-sexing German." In 1980, they were accepted into the Germany Poultry Club. That's where the breed standard comes from, and also the point in time when their name was officially changed to Bielefelder Kennhuhn.

In 2011, Greenfire Farms in Florida imported the first Bielefelders to the United States. They imported unrelated groups of Bielefelders again in 2013 and 2017. Thanks to their efforts, this breed is fairly easy to find across the United States.

## What We Love About Them

Bielefelders are gentle birds, good for families, and beautiful additions to a Chicken Lady or Gent's flock. They do make a good homestead bird with their large frames and good laying ability. While they're calm and friendly, they're still good foragers, and appreciate some supervised free-range time. Their barring gives them some camouflage from predators, but don't rely on this; reports say they're not particularly predator-savvy. They are cold-hardy because of their body mass, but those combs and wattles will definitely need protection from frostbite. They don't love the heat, but they can tolerate it with shade, fans, cool water, and the usual precautions. Bielefelders are one of the few auto-sexing heritage breeds; males and females have different colors/patterns at birth. If you absolutely cannot have roosters, Bielefelders are a fantastic option.

CHAPTER 7

# CHICKS!

Unless you skipped right to this chapter, you've probably done a lot of thinking and planning by this point. You know where your chickens are going to live, what to do with all of those fresh eggs, and maybe you have a rough idea of the breeds that you'd like. But there's more planning you'll need to do. Unless you want to start with point-of-lay pullets, which we discuss later in this chapter, you'll have to do a bit more preparation as you get ready to buy your chicks.

Some chicken keepers absolutely love the chick stage. It is a lot of fun taking care of babies. They're absolutely adorable, and watching them grow up is an experience not to be missed. This is also one of the best opportunities you'll have to handle the chicks and start building a bond with them.

Chick brooding is another great opportunity to get children involved. They can help fill the chick feeder with scoops of feed, add bedding to the brooder after cleanup, and can even hold and pet chicks if very closely supervised. The safest way for a young child to hold a chick is seated (close to the ground), cross-legged for stability, and using both hands to gently support the chick.

## Where to Get Chicks

There are a variety of places where you can source your chicks. Each one offers unique advantages, and the challenge is figuring out which option is best for your purposes.

### Hatcheries

Ordering from a hatchery gives you the best chance of getting the breeds that you want. Some of them specialize in heritage- and rare-breed chickens. Hatcheries have breeding stock (parent birds) on farms in their geographical area; they collect these

eggs and bring them to the hatchery to be placed in incubators. Once the chicks hatch and are dry, they are shipped out via the United States Post Office. Hatcheries offer important vaccines that can be administered at one day old, and they also sex the chicks so that you can order males or females, or both. Reputable hatcheries are part of the National Poultry Improvement Plan, guaranteeing that the birds are tested regularly and found to be free of certain poultry diseases.

## Farm Supply Stores

Buying your chicks from a farm supply store means you can bring them home that day and put them straight into your brooder. The chicks still come from hatcheries and they are still shipped through the mail, but the store receives and unpacks them and deals with any shipping issues. Stores usually have only a few breeds at a time, and sometimes they sell out very quickly. Chicks may or may not be vaccinated, depending on store policy. You may have to ask staff if vaccines were administered at the hatchery. Most stores offer chicks as pullets (sexed female chicks) or straight run (unsexed).

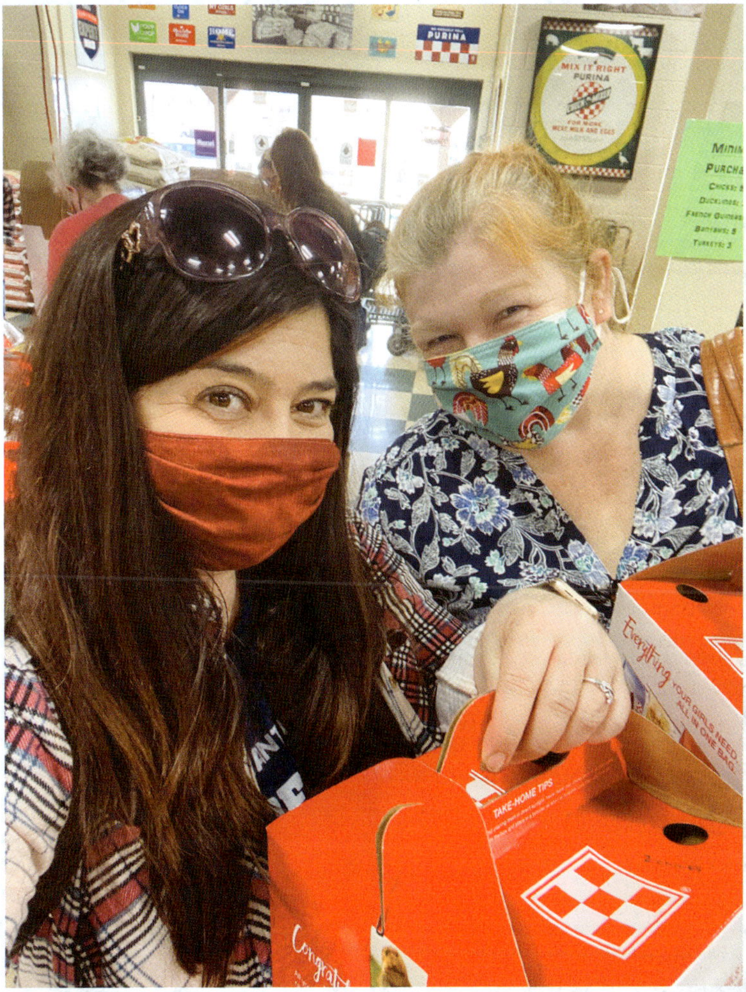


*March, 2021: We braved the crowds in search of Speckled Sussex. Victory!*

### Breeders

Buying chicks from a breeder allows you to bring them home the same day and put them in your brooder. Breeders often specialize in show-quality chickens, so you may want to buy from a breeder if you're interested in showing or breeding chickens yourself. Many breeders do not vaccinate their chicks, and they may not sex them, either. Before you buy from a breeder, ask if they have been certified by the National Poultry Improvement Plan.

## Sexed vs. Straight Run

When you buy chicks at a farm supply store, you'll most likely see a sign on the bin that reads either "pullets" or "straight run." *Pullet* is the term for an immature female chicken, a *cockerel* is an immature male, and *straight run* means unsexed. If you do not want roosters, then you do not want straight run chicks. Also, be aware that even the best chick sexers get it wrong occasionally. Hatcheries have around a 90 percent accuracy rate when it comes to chick sexing. (Chicks are vent-sexed, a visual process performed by highly trained people.) This is the reason why we suggest choosing auto-sexing heritage breeds like the Legbar and the Bielefelder if you absolutely cannot have roosters. Bantams and some of the more delicate, small-bodied heritage breeds are often sold as straight run because sexing them is difficult, and can be very stressful for the chick.

*Little Man, a Polish rooster as a chick.*

## Shipping

*Truffle, a Barnevelder chick, traveling home.*

Shipping chicks via the US Postal Service started in 1918. It is an excellent way to keep rare breeds and their important genetics available to chicken keepers. But how is shipping safe or even humane? When a chick hatches, it's still full from the egg yolk it ingested in the shell and can easily go as long as seventy-two hours without eating. However, most hatcheries do offer nutritional gel to help with the journey. Heat is the primary need of chicks. At the hatchery, day-old chicks are placed in sturdy boxes with bedding and a heat pack, and clearly marked as live animals. When they arrive at your post office, you'll get a call to come and pick them up! They just spent a long time in their egg being rotated (by the hen, or manually/mechanically in the incubator) as they developed, and as long as they are warm and cuddled up with their friends, they should come through the shipping without any problems. Hatcheries will ship with minimum numbers of chicks, often depending on date/weather. While it is not the norm, sometimes there are problems with shipping (e.g., mishandling packages, delays, and unfavorable weather conditions).

Whether your chicks were shipped or you purchased them, you will treat them the same way when you get them home. Have the brooder set up, warmed, and waiting, with good-quality chick starter crumble and water in place. You will take each chick out of the box, and as you set them in the brooder, dip their beak in the water and show them the food dish. They should be eating and drinking around ten minutes after you've put them in the brooder. Keep a close eye on any chicks who are not eating and drinking right away.

## Point-of-Lay Pullets

Some hatcheries and breeders will sell "point-of-lay" pullets. This means they are nearing sexual maturity and will soon be laying eggs. This is a great alternative for someone who does not want to brood chicks or cannot take the chance of getting a rooster. These pullets cost a lot more to buy and ship, but you'll be getting eggs in weeks instead of months. Breeders sometimes call these older birds "grow outs." One downside: It is more difficult to create a bond with these older girls. It is not impossible, but it will require a lot of handling and extra time together.

## Hatching Eggs

Another option for chicks is hatching eggs. You can buy fertile eggs from a breeder and hatch them yourself in an incubator or under a broody hen. There are several things to beware of if you decide to go this route. Make sure you buy from a reputable breeder, or you risk a low hatch rate, chicks with genetic problems, and the possibility of bringing disease into your flock. We've heard plenty of stories about people who thought they were buying one breed, and a completely different breed hatched. You also need a Rooster Plan—see chapter 4—if you are going to hatch eggs. Most clutches hatch 60 percent male and 40 percent female. You will hatch cockerels, and you'll need to know what to do with them.

It is wonderful to hatch eggs with children. They learn so much, and it's fascinating to watch a chick unzip an egg and work its way out. But you should be realistic about whether the risks outweigh the positives.

You may decide to use a broody hen to hatch and raise your eggs instead. Many breeds excel at this, including Orpingtons, Cochins, Brahmas, Salmon Faverolles, and most of the bantam breeds. Some experienced chicken keepers will also let their broody hens raise day-old chicks. This can be tricky if you don't know what you're doing. If you want to try this method, find yourself a good mentor who has lots of

*Martha Washington, a broody Nankin hen, sitting on eggs.*

experience and can show you how to do it safely and effectively. We urge caution here, because if things go badly, chicks can be injured or even killed.

## Vaccines

The two most important vaccines available for day-old chicks are for Marek's disease and coccidiosis. Marek's disease is a virus that attacks the bird's immune system in numerous ways and results in death. Coccidiosis is a protozoan disease that causes damage to the chicken's intestines and can lead to death. Of the two, the Marek's vaccine is arguably the more important, as there is no way to treat the disease, and it can infect your entire flock. Once it is present in the environment, it is very hard to eradicate. Coccidiosis can be controlled by using medicated chick starter feed, or if an outbreak occurs, CORID (amprolium) can be used to successfully treat an outbreak if you catch it early enough.

The vaccines are very inexpensive, usually a dollar or two at the most, and are far more cost-effective than buying remedies for treatment or paying for a vet visit. If you choose to have your chicks vaccinated for coccidiosis, you should not feed them medicated chick feed—it can cancel out the vaccine's effectiveness. Chicks vaccinated for Marek's disease need around a month to develop immunity, so you'll want to keep them away from any older birds that are potential carriers until they're a bit older.

Testing has shown the safety of these vaccines. When the Marek's vaccine was first introduced, there was concern that it was a "leaky" vaccine, and led people to fear that the vaccine would infect their birds with the virus. Most Marek's vaccines now administered in the United States use Herpesvirus of Turkeys (HVT), also known as *Meleagrid alphaherpesvirus* (serotype 3). This strain of the vaccine provides strong immunity but does not cause disease in chickens, and cannot cause contagion. Given the strong science supporting the safe use of this vaccine, and the protection it provides to our own healthy chickens, we do not see a downside to its use.

## Brooders

A brooder is simply a place to "brood" and raise your chicks. It is a structure in which the chicks are kept safely inside with constant access to the appropriate level of heat that they need for their age. They have 24/7 access to food and water, and are safe from any types of predators.

You have a lot of different options for brooders. Anything from large plastic storage totes (around 80 gallons) to soft-sided puppy playpens can be set up as very effective nurseries for your babies. Whatever you choose to use for a brooder must be able to safely contain the babies and provide a barrier from predators, pets, and unsupervised children. The brooder also must contain a heat source, and must be able to retain the heat. In other words, the brooder needs to have sides and a cover. Puppy playpens have mesh zippered tops, while one of the best covers for a storage tote is wire. Chicken wire is light, easy to move, and you can see through it, which makes it a great top.

If you want a more fancy type of brooder, you can make one yourself. There are plenty of plans online if you're interested in building from wood. You can also buy

wooden brooders. This style can be expensive, but if you'll be using it year after year, it will pay off in the long run. Some brooders are basically a wooden box, while others are raised on legs or stands and have built-in places to safely hold heat lamps in place.

You have a couple of options when it comes to bedding for your brooder. We swear by soft wood shavings—not cedar, as it will damage the chicks' fragile respiratory systems—that we purchase from our local farm supply store. Wood shavings are soft, absorbent, insulating, and inexpensive. Hemp bedding is also a popular choice. Hemp is very absorbent, soft, and insulating, but does cost more than wood shavings. Both hemp and shavings provide a good supportive base for a growing chick's feet and legs.

We do not recommend layering newspaper, puppy training pads, or paper towels in your brooder. Newspapers can be slippery and lead to splay leg or other injuries. None of these options absorb droppings effectively. Chicks will poop and spread it all

◀ *Rubbermaid tote and soft-sided pop-up brooder.*

over the brooder as they run around. Fecal contamination can lead to serious illness, including coccidiosis.

## Heat Sources

Young chicks are unable to regulate their own body temperature for several weeks, making them vulnerable to chilling and hypothermia. Without a broody hen to warm them, they rely on the heat methods that you supply for them. Chicks need temperatures of 90–95 degrees Fahrenheit (32.2–35 degrees Celsius) during their first week, but if they have arrived chilled from shipping, they may need even higher heat to revive them. For the backyard chicken keeper, there are two options for heat sources: heat lamps, which have been used for more than a hundred years; or brooder plates, which rely on the same radiant heat technology as coop heaters. Both are safe if used correctly, and can even be used together in certain circumstances.

Of the two types, heat lamps are capable of the highest temperatures. They are the less expensive of the two, and they work in both heated inside rooms and unheated outside spaces, like barns and garages. Because they put off very high heat, they are also a fire hazard. They need to be used carefully; never hang them haphazardly or in a position where the chicks can reach them. If using a heat lamp, you need to make sure the bulb and socket are clean and free of dust and shavings, and that these pieces

Week One: 95 degrees Fahrenheit (35 degrees Celsius)

Week Two: 90 degrees F (32.2 degrees C)

Week Three: 85 degrees F (29.4 degrees C)

Week Four: 80 degrees F (26.6 degrees C)

Week Five: 75 degrees F (23.8 degrees C)

Week Six: 70 degrees F (21.1 degrees C)

Week Seven: 65 degrees F (18.3 degrees C)

Week Eight: 60 degrees F (15.5 degrees C)

cannot come in contact with anything else. The heat lamp also puts out light twenty-four hours a day. Most chicken keepers use a red bulb in their lamp because it is believed to keep the chicks calmer.

Brooder plates are a wonderful addition to the chick keeper's toolbox. Because they use radiant technology, they are zero clearance—they can sit in the shavings and touch everything, including the chicks, without starting a fire. The plate is on adjustable legs, and the chicks duck underneath to get warm. These plates are electric, so just as you would with any other electrical appliance, use extension cords correctly, and periodically check outlets and cords. Because there is no light involved, many chicken keepers prefer the plates to allow chicks to develop their natural circadian rhythm and sleep through the night. The downside is that the plates do not get hot enough to keep chicks warm in an outside space. They were developed to be used in houses and heated rooms. You can lose chicks to hypothermia if their only heat source in a very cold space is a brooder plate.

Whichever heating option you use, make sure there is space in the brooder for the chicks to get away from the heat if they are too hot. This is especially important with the heat lamp. We recommend getting a thermometer or an instant-read heat gun to make sure the heat is at appropriate levels for the chicks. A chick that is too hot may be panting with an open mouth and will lay as far as possible from the heat source. Cold chicks will huddle together for warmth and peep loudly. Very cold chicks may stay in one place with their eyes closed, sometimes not even eating or drinking.

You can start your chicks with both heat sources, and switch off to just the brooder plate as they get older and their heat requirements are lower. Because chicks cannot do without heat, take special care to have backup plans if you lose power for an extended time. For example, take the chicks to a nearby friend's house, or keep a charged battery pack, like a Jackery system, on hand for emergencies. If you are using a heat lamp, buy an extra bulb or two in case the one in use burns out.

Many people take their chicks off of heat at eight weeks of age. While this is fine in temperate climates, it does not work for everyone. If it is still winter, or early spring, your babies may need to have a heat source for longer than eight weeks. In this case, the brooder plate or even a radiant heater like the Cozy Coop heater is a great option. The chicks can run under the plate or stay near the heater if they're chilly. People also sometimes choose to move their chicks outside full-time at eight weeks of age. If you do, we strongly suggest that you give them a heat source so they have the option of warming up when necessary.

## Feed and Nutrition

Nutrition is important for your birds at any age, but it is especially important for growing chicks. Buy them the best chick starter feed that you can afford, and invest in a few packets of vitamins and electrolytes. You can give them a boost in their water anytime they're stressed or growing in feathers. If you're wondering what we feed our chicks, see the Resources section at the back of this book.

### Chick Crumble

Commercial feed companies have developed chick crumble feed that is specially formulated to meet the nutritional needs of growing chicks. The vitamins and minerals

used in chick feed can vary substantially from layer feeds. Chicks don't need as much calcium as laying hens, so chick crumble usually has about 1 percent calcium—enough to help growing bones, but not so much that it damages their kidneys or causes other problems. Chick crumble is usually quite fine so that chicks have no trouble eating it with their tiny beaks.

## Medicated vs. Non-Medicated Feed

Medicated feed contains amprolium, a thiamine (vitamin B1) blocker. Despite the name, this is not an antibiotic. Amprolium is used to prevent coccidiosis outbreaks. Coccidia is a protozoa that attacks the lining of the chicken's intestines, causing watery, sometimes bloody diarrhea. The disease progresses quickly, and chicks can die if they don't receive timely treatment. Coccidia needs thiamine to survive; the amprolium blocks the protozoa's ability to absorb thiamine, effectively killing it. Feeding

*Croissant eating.*

medicated feed gives your chicks a layer of protection against a coccidiosis outbreak. However, you still need to clean the brooder regularly and practice good hygiene to help prevent coccidiosis.

Non-medicated feed does not contain amprolium—it is simply chick food. We recommend that beginning chicken keepers use medicated feed. If you choose not to use medicated feed, we highly recommend keeping CORID on hand to quickly fight any coccidiosis outbreaks. But remember, if you have had your chicks vaccinated for coccidiosis, you should only feed non-medicated chick crumble.

## Chick Grit

Chick grit is simply a much smaller chick-sized version of the crushed granite grit that we feed adult chickens. It isn't necessary to offer grit to chicks if the only thing that they're consuming is chick crumble. But if they have access to grass or any other kind of food, they should have access to chick grit to help them grind the grass, etc., in their baby gizzards. Holly offers chick grit from the beginning so that the chicks are used to it.

## Chick Treats

You can offer treats to your chicks from the beginning, though most chicken keepers usually wait a few weeks until the chicks have grown a bit. If the chicks were raised by a broody hen, she would have them out on grass right away, but she would also be there to guide them. Treats should be quite small at first: tiny snips of grass, very small pieces of berries, or even commercial chick treats. You should always have chick grit available if you are offering treats to your chicks. You don't want your chicks to fill up on treats, so it may be a good idea to offer treats in the afternoon, giving the chicks most of the day to fill up on their chick ration.

# First Aid for Chicks

Chicks are surprisingly resilient and healthy little creatures for the most part. But every now and again, you'll run into a problem. Often, you can get them through it with a little dedicated nursing and a well-stocked first-aid kit. In the photo on page 177, you'll see our chick first-aid kit contents. Most of these items are inexpensive and easy to find.

The chick first-aid kit does have some overlap with the general chicken first-aid kit (see chapter 9). But the items we've shown in the photo are specific to chick care, and when you need them, you really need them. Also, these emergencies tend to happen on a Saturday night or a Sunday morning, when there are no vets open and no one around to help. Be prepared, not scared. Let's take a look at when you might need these emergency items.

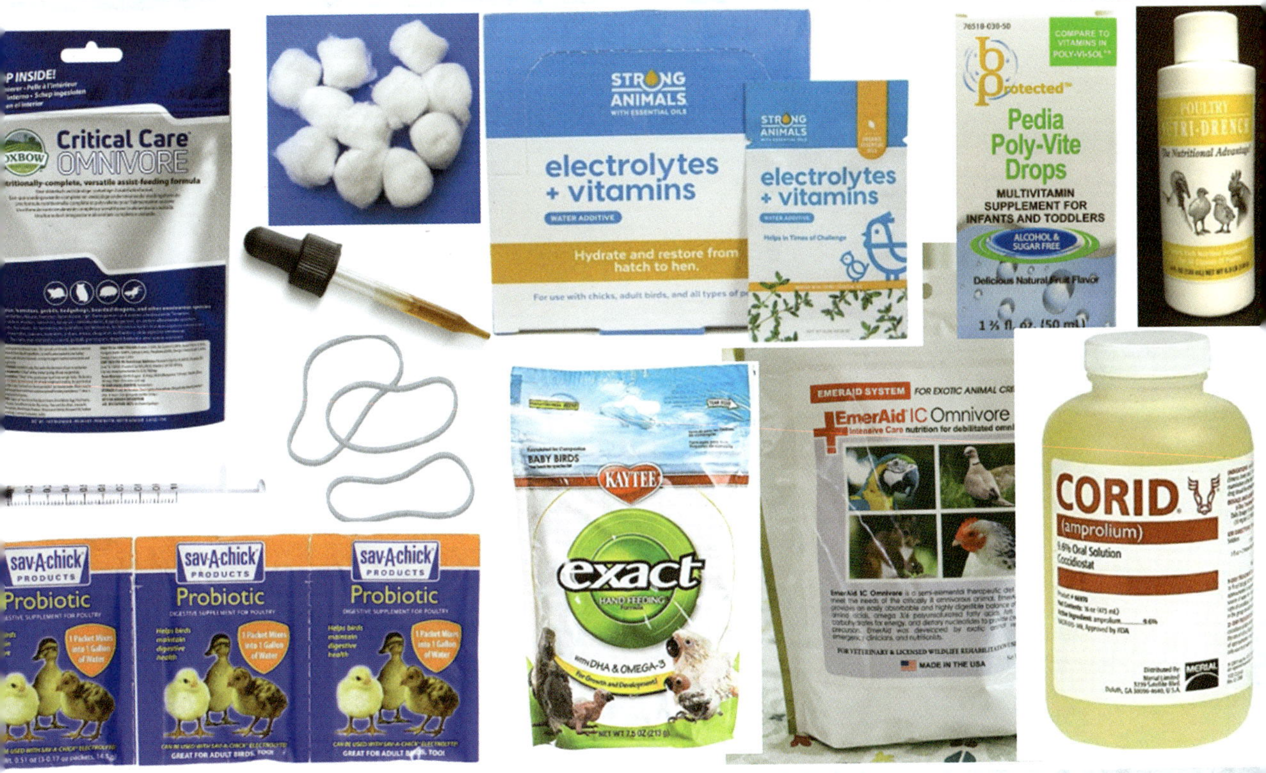

*Chick first-aid kit.*

## Shipping Stress

Chicks who had a rough journey during shipping may need some assistance getting warm, and with eating and drinking. They may stand in one spot with their eyes closed, or they may sleep a lot and not get anything to eat or drink, and you may observe shivering. The most important thing is to make sure they are warm. If you suspect that the baby is chilled, you can gently warm them with a hairdryer on the low setting. Keep a hand next to the chick to make sure the air is not too hot or too strong.

If they are not eating and drinking, use a dropper and place a small drop of Poultry Nutri-Drench against the side of the chick's beak. The chick will usually swallow the drop, but you can try it a couple more times if the chick shakes off the liquid. Nutri-Drench contains electrolytes, vitamins, and minerals at concentrated levels. The chick's body begins absorbing the Nutri-Drench from the mouth/crop as soon as it

*Zara, an Egyptian Fayoumi chick, after a warm blow-dry.*

is swallowed. It stimulates their appetite as well as provides their body nutritionally supportive care. Do not give young chicks more than two drops per day.

## Hypothermia

Hypothermia is the condition in which a chick's core temperature drops to dangerously low levels that may lead to organ shutdown and death. If you notice chicks in the brooder peeping loudly and huddled together, they may be cold. Check the temperature in the brooder with a thermometer or heat gun and adjust accordingly. Individual chicks suffering from acute hypothermia can be gently rewarmed with a hairdryer on the low setting.

## Splay Leg

Splay leg or spraddle leg is a condition where one or both of the chick's legs splays out to the side, making it difficult for the chick to stand and walk. It can be caused by injury during hatching or shipping, slippery surfaces, or inadequate bedding. Fortunately, it is relatively easy to correct. Take a small rubber band or hair elastic from your first-aid kit, hold it lengthwise so the sides are touching, and wrap a piece of scotch tape around the center. Slip the two outside loops around the chick's ankles. This acts as a hobble to hold the legs in the proper position. The chick should be able to move around, eat, and drink. It usually takes just a few days for the problem to correct itself.

## Failure to Thrive

*Failure to thrive* is the general term used to describe a chick that is not growing or that is declining in overall health. Causes may include injury, illness, vitamin and mineral deficiencies, bullying, or genetic issues. Sometimes it is very difficult to tell what has caused an individual chick to experience failure to thrive, so we recommend the same treatment protocol for all: supplementing with vitamins and minerals, such as Poly-Vi-Sol; syringe feedings with powder-to-liquid food; separating from the flock (with a friend to minimize stress) if necessary; and intensive nursing.

*Clover suffered from failure to thrive; intensive nursing saved her life.*

## Pasty Butt

Pasty butt is a condition where feces have stuck to a chick's downy feathers. The feces can quickly build up to the point that their vent is blocked and they cannot poop. This results in pain for the chick and death if the blockage is in place for too long. Use a cotton ball soaked with warm water to soften the dried poop and gently remove it. Do not tug the feathers because the skin is very fragile and can tear. Repeat with the cotton ball swab as necessary.

Causes of pasty butt include brooder temperatures being too hot or too cold, digestive stress causing soft stool, and parasites in the case of chicks being raised by a broody hen. Check brooder temperature and adjust accordingly, feed good-quality chick starter, and make food changes slowly. Offering crumbled quick oats on top of their regular feed has been known to help firm up stool and prevent pasty butt.

Do not put your chick with pasty butt into a glass of water. While it's a popular trend on social media, it is also a recipe for hypothermia and extreme stress for your chick.

## Coccidiosis

Coccidia are parasitic protozoa, single-celled organisms that can infect your chicks through exposure to fecal matter. These protozoa thrive in warm, moist areas such as wet bedding. Chicks who have ingested coccidia develop watery diarrhea that may become bloody as the parasites cause lesions on their intestines. Prevent outbreaks by keeping brooders clean and dry and not overcrowding. Consider feeding medicated feed or getting the coccidiosis vaccine. Outbreaks must be treated quickly with CORID or another form of amprolium.

## Wry Neck

Wry neck is a condition caused by vitamin and mineral deficiencies. The most common deficiency is vitamin E, but selenium and vitamin B deficiencies can also play a role. A chick or young chicken with wry neck will have their neck twist in such a way that they

Silkies are prone to wry neck. If you plan to keep Silkies or other ultra-fancy breeds, make sure they are eating a very high-quality food that meets all of their recommended dietary needs.

are either looking upwards or downwards. This is often referred to as "stargazing." We recommend treatment with Poly-Vi-Sol infant vitamins—without iron, which can damage the liver and cause serious problems—at a dosage of 5 drops (0.1cc) twice per day. Do not overdose with vitamins.

### Vitamin and Mineral Deficiencies

Chicks can hatch with a variety of vitamin and mineral deficiencies, especially if the hen who produced them was deficient. This can cause a number of issues, including wry neck and failure to thrive. One of the easiest ways to combat this is to add vitamin and electrolyte packets to your chicks' water, following the manufacturer's instructions. Offer individual chicks Poultry Nutri-Drench or Poly-Vi-Sol if you suspect additional deficiencies. Do not overdose with vitamins.

## Supplemental Nutrition

In cases where chicks (or chickens!) aren't eating enough during an illness or other stressful period, it can be extremely helpful to offer them supplemental nutrition. This usually comes in the form of powder-to-liquid food, given by syringe. We list our preferred brands in the Resources section at the back of this book. Always follow your veterinarian's or the manufacturer's instructions. If you are inexperienced with syringe feeding, you will want to go slowly. Open the bird's beak, release a small amount of food at a time into the front of the bottom beak, and let them swallow completely before giving any more. Use a 1cc (tuberculin) dosing syringe for best results.

## Handling Chicks

If you're trying to build a strong bond with your chicks, handle them as much as possible. Chicks should always be handled gently and carefully, and children should be under constant supervision when holding babies. Not all breeds will grow up to be super cuddly chickens, but all of that early handling will make it much easier to medicate or otherwise treat your

Chicks start growing feathers to replace their fluffy down as soon as they hatch. By the third or fourth day, you'll see wing feathers starting to sprout. By the end of a week, there will be a little fringe of tail feathers showing.

chickens. For very young chicks, pay close attention to make sure they're not being stressed or getting chilled during periods of handling.

## Chicks Outside

Unless your chicks are being raised by a broody hen, or it's a constant 90 degrees where you live, we do not recommend taking babies outside for the first couple of

weeks. They cannot regulate their body temperatures and become chilled very easily. Here in Maryland, we generally start taking our chicks out for short periods of time at three weeks of age. You'll want to keep an eye on the sky, because predators, including hawks and crows, will swoop down to grab very small chicks, even when you are right there with them. Remember to offer chick grit if your babies have been eating grass or other plants.

If you are a first-time chicken keeper, you won't need to worry about integrating your chicks into a flock—yet. That's for your next batch of chicks. Your babies will simply move out to their new coop and run when they are old enough. Because we have a lot of chilly spring nights, we don't put our chicks outside permanently until they are between twelve and sixteen weeks. If we do put them out earlier, it is with a brooder plate in their coop. Some chicken keepers put their chicks out as early as six to eight weeks. At six weeks, babies usually aren't fully feathered yet. We are very conservative about this because we want our chickens to live long, healthy lives. They may survive early cold stress, but it isn't good for them, and surviving isn't thriving.

## Integration

We swear by our method for drama-free, peaceful, bloodless integration. But here's the thing: With our method, you will not be fully integrating your youngsters until they are sixteen weeks of age. We've heard of chicks injured in horrific ways by older hens, including being literally scalped. It is not uncommon for chicks to be killed by much older, bigger hens. In our opinion, the convenience of having chicks out of the brooder quickly is not worth injury and death.

At around twelve weeks of age, we set up a partition in the run where the new kids will be going to live. We take the young ones out every day, weather permitting. The old guard can see them and even interact with them, but from a safe distance, and supervised, of course. Free-ranging puts them in neutral territory together.

Once the youngsters are just about equal in size (breed permitting) to the originals, we start letting them be together in the run. There may be a bit of minor chasing—and a bit of kerfuffle at bedtime, as they sort out sleeping spots—but there shouldn't be anything more drastic. Slow, easy, low-stress.

If you are having problems, stop and observe the flock for a while. Is there one troublemaker causing all of the ruckus? That bird may need to be removed for a time-out while the pecking order reshuffles. Or do you have mixed breeds that just don't get along? Are your pullets actually younger than sixteen weeks? If they are, back things up for a week or two before you try putting everyone together again. Giving your flock integration a hard reset can make a world of difference. You can get your birds to live happily together if you slow down, practice patience, and let everyone get to know each other better.

# CHAPTER 8

## Feeding Your Birds

Feeding your chickens doesn't have to be complicated and confusing. If you know what they require, you can check feed labels and make an educated decision about what is best for your flock. A chicken's nutritional needs change throughout the different stages of their life, so it is best to feed them an age-appropriate ration that nourishes them and meets all of their dietary requirements. We find it very satisfying to watch our flock devour their food, knowing that it has everything they need to keep them healthy and happy.

On average, chickens eat around a quarter-pound of food per day, though this can be as high as a half-pound, depending on breed, age, and time of year. A quarter-pound equals 4 ounces, or a little over 100 grams. You can weigh it on a kitchen scale if you want to know what 4 ounces of your feed looks like. Chickens should have constant access to feed and water in their run. They do not need food and water in their coop overnight; they'll be sleeping through the night, and the food can draw rodents and other pests.

Water is the most important nutrient that your chickens need. Like food, chickens should have access to clean, fresh water in their run at all times. If your chickens are free-ranging, it's a good idea to have water nearby for them. Water is necessary for all of their systems to function properly. Dehydration can lead to problems producing and laying eggs, difficulty digesting their feed, and an overall lack of energy and wellness. There are a variety of waterers available on the market, or you can just use a water bowl. The important thing is to ensure an unlimited supply of water. In hot weather, you may need to change/add water more often. In a large flock, or a flock with bullies, make sure you provide multiple water sources so that everyone can drink the necessary amount of water to keep them healthy.

Good nutrition is important at every stage, but it is doubly important to laying hens. When breeding chickens, make sure hens are well nourished and receiving their RDA. Many deficiencies can be passed from the hen to the egg, resulting in weak and ill chicks, or eggs that don't hatch at all.

## Layer Feed

If you have hens that are actively laying eggs, we strongly suggest that you feed them a layer's ration. Domestic hens lay so many eggs over their lifetime that it can negatively affect their health if they don't have appropriate nutritional support. Ideally, hens will be eating feed that contains 3–4 percent calcium and 17–20 percent protein. The highest-quality feeds will also offer a full complement of nutrients and probiotics, and may even include support from essential oils. All chickens, including layers,

need a bit of healthy fat in their diet. This allows them to absorb some of the fat-soluble vitamins, including vitamins A, D, E, and K.

Layer feeds also come in different forms. You will have to choose between pellets, crumble, whole grains, or micro pellets. The only difference is the shape. Chrisie's birds prefer crumble, while Holly's like their pellets. But some foods—such as whole-grain or micro-pellet, for example—are only produced one way.

It's a good idea to be familiar with industry standards when it comes to nutritional requirements for laying hens. The definitive guide to avian nutrition is *Nutrient Requirements of Poultry*, produced by the National Research Council. Currently, the ninth edition is available (1994), and can be accessed on the Merck veterinary website. A tenth edition is under way. If you familiarize yourself with the vitamins and minerals needed for chicken health, it makes reading labels on feed bags a lot easier and more helpful.

Hens need all of the usual vitamins and minerals, including A, E, the B vitamins, K, and most of the usual minerals. The most common vitamin and mineral deficiencies and imbalances in layers are calcium, phosphorus, and vitamin D. These three work in conjunction to regulate egg production, and if there's a shortage of any of them, you'll start to see it in egg abnormalities and reproductive issues. Calcium deficiency can be caused by a vitamin D deficiency, since it's needed to metabolize the calcium. Too little calcium could also be due to too many snacks and not enough balanced layer food, or too much spinach or other foods that contain high amounts of oxalic acid.

Oxalic acid is said to lower a hen's available calcium level and cause soft-shelled eggs. According to a 2011 study by West Texas A&M, the University of Balochistan, Pakistan, and the Ministry of Education in Victoria, Canada, hens given supplemental oxalic acid had a reduction in calcium in their eggshells over an extended period of time. They also found that hens getting the supplement stopped eating as much. The higher the oxalic acid in their diet, the less food they will eat. So, go very easy on foods high in oxalic acid, including spinach, beet greens, buckwheat, and potatoes. Rhubarb leaves are so high in oxalic acid that they will poison any animal or human who eats them.

## Oyster Shell and Grit

Laying hens need a lot of calcium, often more than they get from their layer feed. The best supplement is crushed oyster shell. The oyster shell allows the slow release of calcium into the hen's body. Eggshells do provide calcium, but this calcium moves through the hen's body quickly. Without the sustained levels of calcium provided by oyster shell, your hens may lay soft-shelled eggs, have fewer/weaker contractions to aid in egg laying, and they may draw as much as 10 percent of necessary calcium from their own bones. Oyster shell should be fed free choice; your hens will eat what they need. It is possible for chickens to have too much calcium; however, roosters and young chickens generally don't eat the oyster shell.

*Oyster shell feeder.*

Grit is also an essential component of a chicken's diet. Some new chicken keepers mistakenly think that grit and oyster shell are the same. Grit is crushed granite and cannot be absorbed by the chicken's body. Oyster shell is calcium carbonate, which does break down and is absorbed by the body. Calcium carbonate is not hard enough to act as grit. The chicken's gizzard needs grit to be able to grind food; once food has been broken down by the gizzard, it moves on to the proventriculus and starts being digested. We recommend having free-choice access to grit in your run at all times. Chickens who free-range a lot will pick up small stones to work as grit.

## All-Flock Feeds

All-flock feeds are intended to be fed to mixed flocks, birds of different ages, species, and sexes. They're marketed to poultry keepers who keep all of their birds together and want to feed them all the same way. But all-flock feed lacks a lot of the nutrients that laying hens, waterfowl, and other birds need. Your ducks, geese, and laying hens are not going to thrive on an all-flock feed unless you supplement with more nutrients. At a bare minimum, this would be calcium for laying hens in the form of free-choice oyster shell, and niacin for waterfowl, especially ducks.

Chicken keepers with roosters in their flock often feed all-flock, and then supplement with oyster shell so their laying hens get what they need. Many high-quality chicken feed companies have balanced ratios of vitamins and minerals that allow the rooster's body to safely excrete excess calcium. If you do choose to feed your hens and roosters an all-flock feed, make sure you have plenty of free-choice oyster shell available at all times.

## Treats

It is a lot of fun to feed treats to your chickens. They come running fast when they see you getting close with the snack bag! Treats are also a good tool to use if you're training a chicken or getting them used to you by hand-feeding. But it's important that your flock have eaten their regular food all day to get all of the nutrients that they need. Our best practice is to give out a modest amount of treats in the afternoon or early evening.

Chickens eat insects whenever they can find them, so the one treat we offer year-round is black soldier fly larvae (BSFL), or "grubs"—probably the most nutrient-dense dried insect treat you can buy. They're high in protein, calcium, potassium, phosphorus, and magnesium. In the winter, when there are no crickets left, your flock will be super happy to have some bug treats thrown their way. You can also find dried mealworms, crickets, and beetles for sale in farm supply stores and online.

The easiest and most beneficial way to offer treats to your flock is seasonally. Let the seasons guide you in preparing snacks. In the winter your chickens benefit from carbohydrates, so things like scratch grains and oats are great additions to their regular diet. You can feed oats, raw or cooked, hot or cold, and your chickens will love

*Hideo, an Australorp rooster. His shiny feathers show that he is thriving on an all-flock diet.*

*BSFL grubs are always a popular treat.*

them. Winter is also a good time to offer a bit of extra fat to your birds. A little suet, like the kind sold for birdfeeders, is great for chickens. You can even make your own suet blocks using suet from the butcher, or coconut oil.

In spring, as the days get longer and slowly get warmer, your girls are starting to come back into lay. You can begin to taper back on scratch grains and use oats in smaller amounts. Fresh grass and fresh greens are becoming available, so you can offer your flock some nutritious kale, cress, or any other tender greens. Just remember: Stay away from spinach and beet greens, and never let your chicks eat rhubarb leaves.

By the time summer rolls around and the heat is on, snacks that cool and hydrate are the best. Chickens love berries, especially if they've been chilled first. Apples, grapes, and cucumbers are also great for hydration, but the grapes must be cut in half or even quartered if they're big enough to be a choking hazard. Apples and cucumbers should either be left whole so that the hens can take bites, or cut into bite-sized pieces. Melons are the best choice when you're in the midst of a heat wave. Melons of almost every type are about 90 percent water and contain lots of electrolytes, as well as other nutrients. Recent studies have found that lycopene, an antioxidant found in watermelon, cantaloupe, and some other melons, can help reduce the risk of heat stress in poultry.

*Suet cake hanger.*

The cool nights of fall have us adding oats and scratch back into the snack bags. Our girls are starting to molt and benefit from those carbohydrates on chilly evenings. Higher-protein treats are helpful at this time of year, when your birds are growing in new feathers to keep them warm in winter. Peas, grubs, and even dried shrimp are good choices. In severe cases, where a molting bird is not doing well and not eating enough, you can offer them a bit of tuna. This is an old-timer's trick and an excellent way to get protein into a struggling hen, fast.

## Nutritional Supplements

Ideally, your flock is getting all of their nutrients from their diet. But in times of stress or illness, it can be extremely helpful to offer your birds a boost. These days, there are plenty of options on the market for probiotics, vitamins and minerals, and electrolytes.

Apple cider vinegar, better known as ACV, works well as an occasional water additive. It helps acidify the chicken crop, making it harder for bacteria and yeast to overgrow. ACV should be added to your chickens' water at a formula of 1/1/1: one tablespoon in one gallon of water, once per week. ACV is not a cure-all and cannot work in place of an antibiotic, anti-yeast medication, or dewormer. Too much ACV can cause digestive upset and diarrhea in chickens.

Many of these options are liquid or powder water additives, though some are powder-based feed additives. Of these two methods, we find the water additives to be much more effective.

When using these supplements, always read and follow the manufacturer's instructions. These products all have their own doses and frequencies, meant to be used as directed. It is possible to overdose a chicken with nutritional supplements, especially vitamins and minerals. There are a few stronger supplements that are given to sickly birds by beak, and we will cover those in chapter 9.

## Kitchen Scraps

For the most part, leftovers and kitchen scraps are okay to feed your chickens. They should not be your flock's only diet; that would leave your birds open to all kinds of vitamin and mineral deficiencies and malnutrition. It is very popular for chicken keepers to claim that their grandmother only fed her chickens leftovers, but this is only partially true at best. If it were true, Grandma's chickens probably wouldn't have made it through the winter. There has been plenty of good-quality chicken food on the market for the last 110 years. Prior to that, people fed their flock mixed grains, primarily corn, oats, and barley. Poultry were definitely offered a lot of the scraps and excess produce, too, and the birds supplemented with grass and herbs and plenty of bug catching.

As a general rule when feeding straight from the garden or farm stand, produce should be left whole so that the chickens can take bites from it and not choke, or cut into bite-sized pieces. Large fruits like melons or pumpkins can be roughly cut into quarters and left on the rind.

Never feed your chickens moldy food or spoiled leftovers. This can make them very sick, causing serious crop and intestinal infections, or worse. Moldy food should

*Mary Berry eating salad leftovers.*

be composted or thrown away. Spoiled food is generally contaminated with bacteria and also should be thrown away. You should also avoid feeding long strings of fruit and vegetable peels because they can cause an impaction in the chicken's crop.

Oatmeal is perfectly safe to feed to your chickens, and it's very likely that your flock is already eating oats. Most chicken feed in the United States contains oats, because they are relatively easy to grow and one of the most common cereal crops in both America and the world in general. Oats were a major part of chickens' diet in Colonial and Early America, and millions of chickens have thrived on oat-inclusive diets for thousands of years. In case you need science to back up the history, numerous recent studies have pointed to the positive effects of oats and oat beta-glucans on poultry health.

Foods to avoid feeding your flock: avocado, citrus, onions, raw potatoes, chocolate, alcohol, uncooked dry beans, green tomatoes, raw or undercooked meat, wild mushrooms, fruit pits and seeds, caffeinated foods, fried or salty foods, and highly processed foods.

*Lucy and the girls enjoying a bowl of oatmeal.*

*First-aid supplies.*

CHAPTER 9

# First Aid and Health

When it comes to keeping chickens, prevention is half the battle. Preventive care goes a very long way toward keeping your flock healthy, but eventually injuries, accidents, and illnesses will happen. That's why we encourage you to keep a well-stocked first-aid kit and to learn how to give your chicken a basic health exam. (Don't forget to brush up on your basic chicken anatomy in chapter 3!)

We also strongly encourage you to find a veterinarian, and we'll discuss that in more detail later on in this chapter. Chickens are living creatures just like every other, and they deserve medical treatment when they have problems. Even if you only view them as livestock, they still deserve veterinary care, just like cows, sheep, and any other animal.

## Building a First-Aid Kit

When illness or injuries happen, you want to be prepared. All too often, emergencies happen on weekends or holidays, or other times when you either can't get to your vet or you can't get to the store for supplies. You don't have to run out and buy absolutely everything on this list right away, especially if it doesn't fit into your budget, but it's a good idea to start with a few basics and add more whenever you can. You can buy one item every time you're in the feed store or farm supply store. Several of the items can be ordered from Amazon, some from Chewy, and some of them can be found in your local pharmacy. Start with the essentials, and add next-level supplies when you can.

## First Aid Kit Essentials

- CORID—liquid or powder, treats coccidiosis
- Dewormer—liquid Safe-Guard for Goats
- Vitamins and electrolytes—almost any brand will do
- Poultry Nutri-Drench—a liquid nutritional supplement that helps balance electrolytes, vitamins, and minerals, very quickly—don't be without it
- Vet wrap—self-adhesive bandage, can be cut to size
- Sterile nonstick pads—any size will do; you can cut down the larger ones
- Silver ointment—used as a topical antibiotic and antifungal; if you can't get prescription strength, several over-the-counter versions exist. Use in place of triple antibiotic ointment, especially if you don't know what bacteria/pathogen you're fighting.
- Tripe antibiotic ointment—only use the version without pain medication
- Vetericyn wound cleaner
- Syringes—1cc/ml and 3cc/ml
- Nolvasan / Hibiclens / Iodine—topical antimicrobial (bacteria and fungus) and antiseptic cleanser or scrub
- Blood stop powder
- Terramycin eye ointment
- Petroleum jelly or Vaseline
- Sterile gloves
- Headlamp
- Epsom salts—for loosening bumblefoot scabs, and as a therapeutic soak for sprains and strains and lameness issues

### Next-Level Supplies

- Poly-Vi-Sol—liquid vitamins, especially helpful for chicks and/or treating wry neck
- Bent-tip glass dropper—the angled tip can be placed against the beak and makes dosing much easier
- Omnivore Care (from a vet) or Kaytee Baby Bird Formula—powder-to-liquid food, for chickens who are compromised, too ill to eat, or recovering from crop surgery
- Crop bra—we strongly recommend the Birdy Bra Crop Supporter from Crazy K Farm
- Soft-sided pop-up animal playpen
- ZeroClearance heat panel
- Small nail clippers
- Nail files
- Tweezers
- Ivermectin injectable dewormer—use only as topical treatment for external parasites
- Towels and blankets appropriate for chicken use

*Next-level supplies.*

## Preventive Health Care

What exactly is preventive health care? Essentially, it is providing your chickens with excellent care to keep them healthy and thriving. It includes keeping a very clean coop, feeding fresh, good-quality food, supplying fresh and clean water every day, and maintaining a low-stress environment. Be observant, and try to think ahead about possible problems. For instance, if you want to free-range your birds but the lawn is unmowed and full of long, coarse grass, you may want to keep them inside until the grass has been mowed.

Keep an eye on your flock and know what "normal" looks like. If one of your chickens is getting sick, the sooner you start treatment, the better her chance of recovery. Some chicken keepers like to add herbs to their chickens' feed, or even make "tea" by putting herbs in the water. There's a growing body of scientific evidence that proves

*Pansy, a normal, healthy chicken.*

some herbs can have a beneficial effect on chicken's health, bolstering their immune system and making them more resilient. A side effect of this is good digestive health and good eggshell quality. One of our favorite herbal products is Chicken E-lixir, a blend of herbs and essentials that you add to your flock's drinking water. Oregano is the star of the supplement, along with some other vitamins and minerals, and it does a great job of keeping crops healthy and feathers shiny.

We want to stress that herbs should be used as preventives, not treatments, unless you are very experienced in herb craft and botanical medicine. By the time a chicken is showing signs of illness, they need treatment very quickly, and usually with a pharmaceutical product that is strong and fast. Herbal remedies generally have to be administered several times per day for effectiveness, and it is difficult to know the potency and amount necessary for therapeutic healing. Both under- and overdosing can end in death. When it comes to treating a beloved chicken, we always opt for the solution that will give the best outcome for recovery.

## How to Give Your Chicken a Ten-Point Health Exam

Chickens, like a lot of herd or flock animals, are very stoic about illness and injury. It's a survival mechanism, protecting them from both predators and the rest of their flock. The problem is that the stoicism might be masking clues that we need to see to figure out what's going on with our poultry.

You can give a health exam anytime you want. You may want to do random checks on your birds to make sure they're healthy, but sometimes you'll be giving exams because you see a chicken that isn't acting normally, or you may notice a physical symptom that needs a closer look. We're going to tell you how to give a quick exam to try to eliminate possible health issues. This is also very useful if you're calling a vet for help. They may ask you if you've checked these things. Don't be afraid to revisit chapter 3 for help with chicken anatomy.

1. **Head/Face:** Skin should be bright red and healthy, no swelling, no abrasions, no mites or lice. Combs and wattles should not be pale or dry and shrunken. Ears should be clean with no discharge.
2. **Eyes:** Eyes should be clear with no swelling, cloudiness, or discharge.

3. **Nostrils/Mouth:** Nostrils should be clear with no discharge; tongue should be pink and not pale; mouth should be free of foreign materials or canker sores, scabs, and lesions.
4. **Crop:** The crop should be flat or almost flat in the morning, not full of squishy fluid or firm material. During the day, the crop should feel like it has food in it, and at night should be plump and soft. Some ladies have a big, healthy crop at night, more commonly known as the "food boob."
5. **Keel/Breast:** The chicken's breastbone, or sternum, runs from their chest to their abdomen, and the area is called the keel. The breastbone has a projection that runs along the front of it and can be felt by placing your hand on the bird's chest. There should be padding from the breast muscles, which sit on either side of the bone, and the bone itself should not jut out sharply. Check the keel for any blisters, scabs, or lumps. The keel bone should be well padded with muscle and fat. If the keel bone is sharp or prominent, your bird may be underweight and in need of more nutrition.
6. **Abdomen:** The skin on the abdomen should be tight and firm. Feel the abdomen for any scabs or fluid under the skin.
7. **Under Wings/Feathers:** Check carefully here: Are there any lice or mites? Any wounds? Is the chicken able to use the wing?
8. **Vent:** The vent should be pink and clean. Is there any swelling? Is there any discharge, and what color? Is there poop stuck to the vent? Are there lice and mites or egg casings?
9. **Legs:** Are the scales in the legs smooth and healthy? Any lifted scales or crusty material accumulated around the scales? Are there any visible injuries or lameness? Any swelling or heat in the joints? Any broken toenails? Are toes in the usual position?
10. **Bottoms of Feet:** Foot pads should be soft and healthy. Are there any black scabs or swollen areas? Any visible wounds or inflammation? Are they walking normally, or showing signs of lameness or limping? If your chicken is lying down frequently, or more than usual, check the bottoms of their feet.

## Creating a Chicken Hospital

Birds with minor injuries and noncontagious illness can often stay with the flock, but more serious injuries or illness calls for birds to be safely housed temporarily in their own "hospital" unit. Separating chickens from the flock may be necessary for quarantine to stop the spread of illness, to keep the sick bird safe from bullying and in a lower-stress environment, and to prevent fly strike on wounds and vulnerable chickens. Having a chicken hospital setup also gives you better access for nursing and medicating that bird, and it gives you a chance to observe their behavior, monitor their food and water intake, easily check their poop, and make sure they are comfortable and warm or cool enough, depending on the season.

*Gertie, comfortable in the chicken hospital.*

Your hospital should be safe from the flock, predators, and household pets. Unless the chicken is gravely sick or injured, it's a good idea to make sure they have enough room to move around. Soft wood shavings or hemp are ideal bedding unless the bird has crop problems or recently had surgery. In that case, you'll want to put her on towels (we prefer white, because it's easier to monitor poop or fluids) so she can't eat any of the shavings.

Our favorite hospital unit is a soft-sided pop-up puppy playpen. They're soft; easy to set up, move around, and store; and they have zippered tops and doors for added security. You can even move them outside for supervised outings with your feathered patients. We do know of people who have effectively used camping tents. In a pinch, you can use a large cat carrier or a small dog crate inside a safe, heated, or climate-controlled space. We recommend that you keep some extra bowls on hand, as well as radiant panel heaters, for birds that need help keeping warm, and fans for hot weather. Bonus: These hospital units can also be used in case of severe weather or evacuation. Double bonus: Buy a pet or chicken stroller for a mobile hospital unit. We use ours for birds that need intensive care. Strollers make it easy to keep the chicken with you wherever you are.

## Crop Disorders

A healthy crop has its own balanced micro-flora with a pH that supports good bacteria and digestive health. This is why we offer chickens pre- and probiotics when they have digestive issues, or when they're on antibiotics or other medications that could disturb the harmony of those microbes. Understanding the chicken crop and offering nutritional support and preventive care goes a long way toward helping your chickens maintain healthy digestion.

### Crop Infection

One of the most common ailments of the crop is infection. Chickens can develop bacterial infections, fungal infections, and co-infections of both. Sometimes there is a foul smell from the crop, but this is not always the case. Absence of smell should not be used to dismiss the possibility of crop infection. Crop impaction may occur at the same time.

*Gertie, with crop infection, wearing a DIY crop bra.*

*Common Symptoms of Crop Ailments*

- Crop filled with fluid and feels squishy
- Poor or no appetite
- Crop still full in the morning
- Lethargic behavior
- Hunched posture and puffed/ruffled feathers
- Segregating from the flock
- Regurgitation/fluid leaking from the beak
- Infrequent poop or no poop
- Very watery poop with little to no solids
- Jerking the neck from side to side, trying to move crop contents ("snake-necking")

Crop infections are caused by an overgrowth of bacteria or fungus, or both, often because the chicken has eaten something contaminated. Infections can also occur when the crop contents are not moving, allowing for pathogen overgrowth. In our experience, crop infections are sometimes the result of a larger problem. Parasite loads can cause intestinal blockages that lead to crop infection. Molting or otherwise compromised chickens may be bullied and pushed away from feeders. As a result, they may start eating inappropriate things that are contaminated. Illnesses such as heavy metal poisoning, cancer, and Marek's disease can also cause slow crop and resulting infections. Crop infection can also be caused by mechanical issues, such as crop blockages from long and fibrous grass, table scraps, and straw.

## Crop Impaction

A crop impaction is essentially a blockage of the crop. Chickens, especially hens, may eat coarse plant matter, fibrous fruit or vegetable peels, or even hay and straw that cannot pass through their crop. This material forms a firm mass that effectively plugs the crop and sometimes cannot be resolved without surgery by a licensed veterinarian. We cannot advise you to attempt clearing the crop on your own. It is risky, painful, and can result in the death of your chicken. If left untreated, a crop impaction can lead to a crop infection or other crop problems.

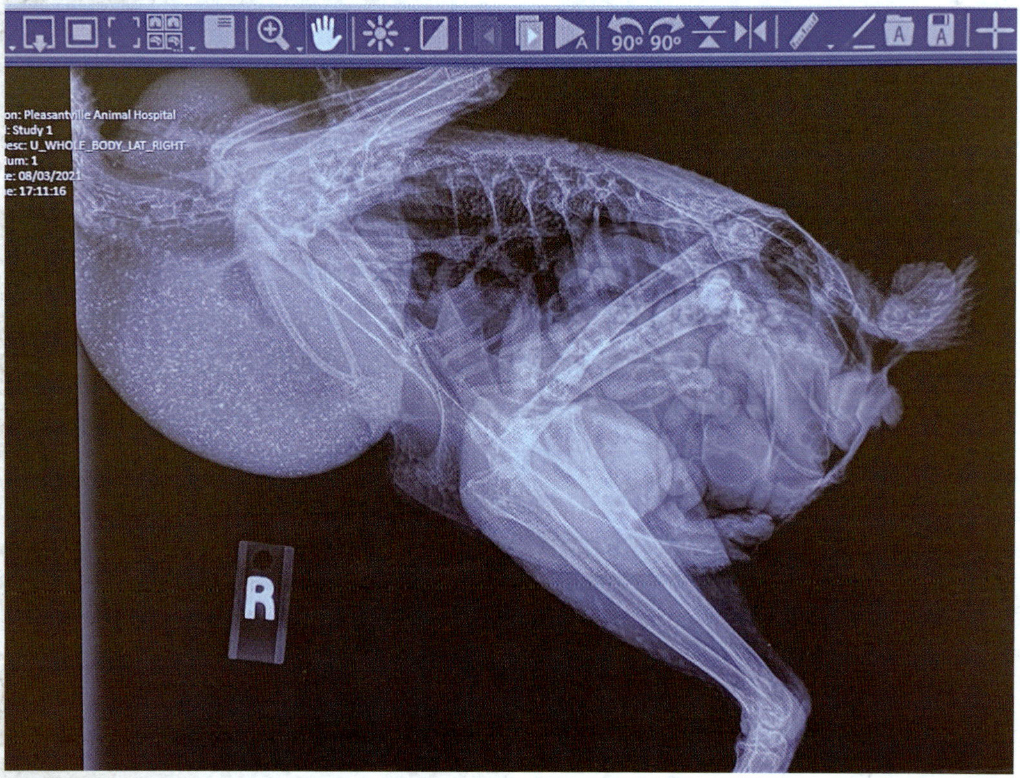


*X-ray of Gertie's full crop.*

## Slow Crop / Crop Stasis

When the chicken crop stops working, this is known as crop stasis. Slow crop is delayed emptying of the crop contents. Both of these conditions can lead to crop infection and can be caused by Marek's and some other poultry diseases, so we advise consulting with your veterinarian if your chicken develops slow crop or crop stasis.

## Treatments

Your first line of treatment for crop problems should be a visit to your veterinarian. Your vet can perform a cytology to diagnose a crop infection. They will take a swab of the chicken's crop and look at it under a microscope to check for the presence of abnormal bacteria and yeast. Depending upon their findings, they will prescribe antibiotics, nystatins (antifungal medication), or both. They may also prescribe motility medications, such as metoclopramide, to help the digestive system move, and subcutaneous fluids to help with dehydration.

Chickens with crop problems often cannot or do not eat and drink enough. It is crucial to offer them nutritional support. Poultry Nutri-Drench is a quick and effective way to balance electrolytes and get vitamins and minerals into their system. It works very quickly because it is absorbed from the mouth and the crop and does not need to be digested. Powder-to-liquid food can also be used to get necessary calories and fluids into a sick bird. You mix the powder with water, following the manufacturer's instructions, then slowly feed it via a dosing syringe. Open your chicken's beak and slowly and carefully squirt small amounts of liquid food into the front of their lower beak. Allow the bird to swallow completely before giving more.

In cases where intestinal parasites are suspected, deworm your chicken with 1cc of liquid Safe-Guard for Goats. That dosage is for standard-sized chickens. The dose for bantams is 0.5cc. There is a ten-day egg withdrawal. It is an off-label use of this dewormer, but highly effective. For more about internal parasites, check the parasite section in this chapter.

## Crop Bras

Bras aren't just for women anymore. Your hen may need a bra if she is struggling with crop issues or if she has had crop surgery. Post-surgery, a crop bra protects stitches and supports the crop muscles as they heal. During a crop infection, a bra can help support the crop in the proper position, allowing it to empty more easily.

You can buy a bra—the absolute best is the Birdy Bra Crop Supporter from Crazy K Farm—or you can make one using vet wrap. See the Resources section for a link to our YouTube tutorial.

For a crop bra to be effective, it needs to be worn properly. The bra straps are tied around your chicken's neck and chest. They should be snugly tied, and the bra itself should hold the crop as if it is on a shelf. If the crop bra isn't lifting and holding, then it will do no good. A properly fitting crop bra can help your hen immensely.

While roosters occasionally have crop problems, it is far more likely for hens to develop crop issues, such as infection and impaction.

*Gertie in her crop bra.*

## Bumblefoot

Sounds funny, but it's not. Bumblefoot is a foot abscess, also known as ulcerative pododermatitis. It frequently occurs in birds and animals kept on wire. Bumblefoot occurs more often in the heaviest chickens, as they are more prone to abrasion and small cuts on the bottoms of their feet. These small injuries allow bacteria to enter the tissue. Splintery roosts can also be a cause of trauma, as can hot and abrasive dry ground in drought conditions.

A simple case of bumblefoot presents as a black scab on the bottom of your poultry's foot. The bird may or may not be limping. Sometimes it is very painful; other times it's minor, and you won't catch it unless you look at the bottom of your chicken's feet. Bumblefoot infections can be more complex and will sometimes move through the foot. Lesions and inflammation can also occur between the toes, on the tops or sides of the toes, and on the top of the foot. Untreated bumblefoot can become very serious, leading to bacteria in the bone and necrotic tissue that may need surgery or amputation. Very serious cases can lead to sepsis and death.

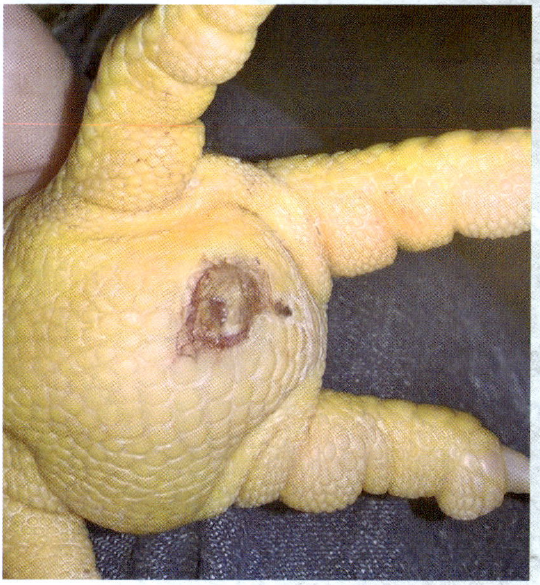

*Kasper had a case of complicated bumblefoot that required surgery.*

### Treatment

Simple bumblefoot can be treated in a straightforward way: First, you will need to remove the scab. Sometimes you can easily peel the scab, and sometimes it needs to be soaked before it will come off. Always use a clean bowl or tub for soaking the foot, and add Epsom salts or chlorhexidine (Nolvasan or Hibiclens) to the water.

The junk underneath the scab can vary in appearance—sometimes liquid, sometimes more solid, sometimes attached to the scab. In some cases you can't get to whatever is inside. Besides soaking, you can try a drawing salve (like Hyland's PRID),

*Drucilla with a bandaged foot.*

or even a plantain poultice (made from the leaves of the *Plantago major* plant), changed regularly. If the infection persists, you should take your chicken to see their vet.

If the infection is local and confined to right under the scab, you can clean and debride the area of any exudate (pus and debris); spray with antiseptic; then apply silver ointment, changing bandages and reapplying ointment as needed until the foot has healed and the scab is gone. We use nonstick pads and strips of vet wrap to bandage our bumblefoot cases. See the Resources section of this book for a link to our bandaging tutorial.

There is an Internet myth that bumblefoot infections are always caused by *Staphylococcus aureus*. This is incorrect. Many different bacteria cause bumblefoot, and yeast/fungus may also be present. Holly's Swedish Flower rooster Kasper had a bumblefoot culture come back from the lab with three different bacteria (including *Klebsiella* and *Enterobacter*) present, plus candida/yeast. Chrisie's Salmon Faverolles hen, Ivy, had a culture come back from the lab showing *E. coli*, *Staphylococcus aureus*, and a gram-negative bacterium. You may want to consider having a culture done if at all possible. Knowing what antibiotics the bacteria are sensitive to lets the doctor prescribe effective medications that may be used both topically and internally.

Most cases of bumblefoot clear up once the exudate is removed and the foot heals. But more complex and stubborn cases may require your veterinarian to perform surgery. We do not recommend DIY bumblefoot surgery. Veterinarians work in a sterile environment with anesthetic and pain medications. Many "home" surgeries cause the chicken a lot of pain and introduce more bacteria. Too many of these home procedures end in the bird's death.

## Prevention

Like most other poultry diseases, preventing bumblefoot is easier than treating it. Protect the feet of bigger and heavier chickens by having lower perches in place to lessen the impact and avoid injury. A thick layer of soft shavings also helps cushion feet. Keep the perches and roosting bars in your coops clean and smooth. You can scrub them with a mixture of one part vinegar to one part water to disinfect, and you can sand your wooden roosting bars two or three times per year.

Be proactive and check the bottoms of feet on any birds that you see limping or lying down frequently. See also the ten-point chicken exam on pages 207–8.

## Respiratory Problems

As we showed in chapter 3, chickens—and birds in general—have a much different respiratory system than mammals. Because of the birds' fixed lungs and nine air sacs, there are several areas that can be compromised (airsacculitis). Chickens normally breathe in through their nostrils, so anytime you see a chicken breathing through their mouth, you may want to check on them and make sure they aren't having breathing problems. Stress and overheating are the most common reasons for open-mouthed breathing, but a bird gasping or wheezing may be the start of a serious problem. A bird presenting a blue or purple comb should be examined by your vet right away. Prevention is the key when it comes to respiratory infections. Please consult with your veterinarian as soon as possible if you suspect respiratory problems in your flock.

### Viruses

Infectious bronchitis virus, avian influenza, Newcastle disease, and infectious laryngotracheitis virus are some of the most common viral respiratory diseases that can affect chickens. They are spread by respiratory droplets and cause coughing, sneezing, eye and nasal discharge, and sinus/facial swelling. While high-pathogenic avian flu is arguably the most serious of these viruses—and in rare cases can infect humans—none of them have effective treatments, and they can lead to high mortality.

### Bacteria

The two most common bacterial infections in chickens are infectious coryza and *Mycoplasma gallisepticum*. Coryza causes inflammation and swelling of the face and sinuses—sometimes severe—and discharge from the nostrils and eyes. Treatment consists of antibiotics and non-steroidal anti-inflammatories, as well as intensive nursing and supportive care. In *Mycoplasma gallisepticum*, coughing, sneezing, nasal discharge, and facial swelling are common, but mild subclinical infections are also possible. Hens infected with mycoplasma often have a noticeable drop in egg production. Treatment includes antibiotics, non-steroidal anti-inflammatories, subcutaneous fluids, and supportive care. While these bacterial infections don't spread to humans, chickens may carry and shed other bacteria that can infect humans.

## Fungus

Aspergillosis, also known as brooder pneumonia, is the most common chicken respiratory disorder caused by fungus. Chicks and immunocompromised birds are more susceptible, and are often infected after breathing in the mold spores in a dirty brooder. Symptoms include labored breathing and gasping for air, as well as reluctance to eat and drink. Treatment involves very lengthy use of antifungal medications and supportive care. Clean, dry bedding is absolutely essential. Aspergillosis may be contagious to humans.

## Injury

Chickens can sustain injuries to their respiratory system, including inhalation of smoke or chemical/toxic fumes, or suffer damage from predator attacks. Birds with

*Drucilla in a support bandage after tearing her ACL.*

these injuries should be examined by your veterinarian right away. Treatment will vary. The vet may prescribe various medications to treat injuries and keep them comfortable. Nursing and supportive care will be necessary.

### Prevention

Keeping your birds in a clean environment is your first line of defense against respiratory problems. Try to keep dust under control to avoid irritating your chicken's respiratory system. Equally important is following strict biosecurity protocol. This includes quarantining new birds for at least six weeks, not buying chickens from people with mycoplasma-infected flocks, avoiding purchasing poultry at auctions or swaps, changing footwear and clothing between visits to other farms and visits to your own birds, and not taking your chickens to places where other poultry are present.

## Internal and External Parasites

At some point in your chicken-keeping adventures, you'll probably come up against some parasites. Parasites are organisms that live off of your chickens, robbing them of nutrients. Parasites are gross and dangerous, and it is a mistake to ignore them. Both external parasites and internal parasites in large numbers can cause your chicken to lose weight quickly and even become anemic. Severe cases, left untreated, may result in death.

The external parasites, including mites and lice, are easy to diagnose with the naked eye; use a headlamp for best results. Internal parasites are harder to diagnose. Only two types of intestinal worms are visible in chicken poop. The rest require a microscope and a slide to search for eggs. This is known as a fecal egg count, or FEC. If you're not equipped to do it yourself—and it is challenging—you can drop off a stool sample at your veterinarian's office. Without an FEC, there is no way to know if parasites are present and/or if your deworming efforts are successful.

### Internal Parasites

Chicken can pick up parasites from a variety of places. Parasite eggs are left behind in the poop of wild birds and other animals. Even the cleanest-looking lawn or chicken run could have microscopic parasite eggs present. Overcrowded coops and runs mean more contact with poop, and a higher risk of passing around parasite eggs. Chickens spend a lot of time eating grass and digging in dirt, leaves, and even coop bedding—all good opportunities to pick up parasites. Some earthworms can pass parasite eggs when chickens eat them.

Bringing in new adult birds can also be a source of parasites. We recommend deworming new adult chickens and quarantining them for at least six weeks, following biosecurity protocols. You may want to have fecal testing done by your veterinarian's office to make sure the newcomers are clean.

*Roundworm visible in chicken poop.*

While not all chickens will show symptoms of a parasite load, there are some warning signs that you should not ignore. Consider treating any birds who show signs of lethargy, pale comb, weight loss, foamy-looking poop, diarrhea and messy butt feathers, decreased egg production, decrease in eating and drinking, ruffled feathers and unthriftiness, and generally looking or acting unwell. Slow crop or other crop ailments may also indicate a parasite load.

In the United States, the most common intestinal chicken parasites are roundworms, tapeworms, cecal worms, capillaria (thread worms), eye worms, and, rarely, gape worms. Most of these parasites can be killed using a broad-spectrum dewormer called fenbendazole. Brand names include Panacur and Safe-Guard. We recommend deworming your chickens orally with 1cc of liquid Safe-Guard for Goats. That dosage is for standard-sized chickens. The dose for bantams is 0.5cc. A second dose at ten days is also recommended. There is a ten-day egg withdrawal. This is an off-label use of this dewormer, but highly effective.

Eye worms, capillaria, and some forms of tapeworm may require different dewormers, and fenbendazole is not always effective against them. If you suspect that your flock has one of these worms and you cannot get it under control, contact your veterinarian. Eye worms are usually only diagnosed by a veterinarian and require eye drops.

## External Parasites

External parasites can be just as gross and just as dangerous as internal worms. Common carriers include other chickens, wild birds, and rodents. Carefully check any new adult chickens for external parasites, and use good biosecurity practices and quarantine them before integrating them with your flock. Unfortunately, chickens adopted from shelters often bring mites along with them. If you are inexperienced with external mites, consider treating the newcomers prophylactically. Be vigilant about wild birds; sometimes they build nests in larger coops, and that can introduce mites and lice to your chickens. Remove any nesting material and spray with a pyrethrin-based product; Elector PSP; or clove, mint, or citrus essential oils.

The most common external parasites are northern fowl mite, common chicken lice, feather chewing mites, scaly leg mites, and, rarely, red mites. While red mites are not common in the United States, they are a problem for our friends in the United Kingdom. All of these pests, except scaly leg mites, are visible to the naked eye. Warning signs include chewed, dull-looking feathers, pale combs, messy vent feathers, feather loss, broken and irritated skin, itching and head shaking, and the appearance of cement-like egg casing around the vent and abdominal feathers. In severe cases you may notice ruffled and hunched birds, lethargy, decrease in eating and drinking, and decreased egg production.

It is fairly easy to spot mites and lice. Arm yourself with a flashlight or, even better, a headlamp to have both hands free. Put your chicken up on a roosting bar or some other surface where you can carefully look her over. Beginning at her vent, shine the light, part her feathers, and look carefully for movement and/or egg casings attached to the feather shafts. Go over her whole body, paying special attention to the warmest spots: the vent, under the wings, at the base of the tail, and the neck feathers. Chicken lice are the largest of the pests and the easiest to spot.

Scaly leg mite is generally spread from chicken to chicken, especially during winter when the birds snuggle together for warmth. The mites will pass from chicken to chicken, generally when the flock is on the roosting bars at night. The smooth scales on the chicken's legs will become raised. In severe cases, crusty material will show under the scales.

*An example of scaly leg mite on a rooster.*

## Treatment

Treatment for mites and lice, including scaly leg mites, is Ivermectin. The most effective form of Ivermectin to use is 1 percent injectable solution. But *do not inject it*—you use it topically. Do not try to use the pour-on form of Ivermectin for this treatment, as it will irritate and injure your chicken's skin. Use a needle and a 1cc syringe to draw up the solution, then remove the needle. The dosage is 0.25cc for a standard-sized hen, 0.3cc for a standard-sized rooster, and 0.125cc for a bantam. Apply the Ivermectin directly to the skin at the base of the chicken's neck, in between the shoulder blades. This is similar to applying topical flea control on a dog. You will treat your chicken three times over the course of a month. The first application is day one, the next application is at fourteen days, and the last application is at twenty-eight days.

For scaly leg mites, it is also beneficial to coat the chicken's legs in petroleum jelly or coconut oil. Gently push the coating up under the scales. This helps smother the mites, and also helps soften the crusty exudate that builds up under the scales. In some cases the raised scales will fall off, but the chicken will grow healthy new ones.

## Prevention

The best way to help prevent mite and lice infestation is to have a nice, big dust-bathing area for your flock. Chickens will pick off the odd critter as they preen themselves, and the dirt and dust from the bath will help smother the parasites as the chickens roll in it. You can add wood ash, diatomaceous earth, and insect-repelling herbs, especially in the mint family. After regular coop cleaning, give the interior and the roosting bars a spray with pyrethrin-based product, or go natural with diluted clove, mint, or citrus essential oils. Elector PSP is also a popular option with some chicken keepers. Allow the surfaces to dry completely before adding bedding or closing in your chickens.

## Reproductive Problems

Because it is rare (but not unheard of) for roosters to suffer from reproductive issues, we are concentrating on hens in this section. Our beloved hens have been bred over hundreds and hundreds of years to maximize egg production, and along with that comes stress on the reproductive system. Such intense production can lead to a host of problems. Some of the more serious issues must be seen by a veterinarian for effective diagnosis and treatment.

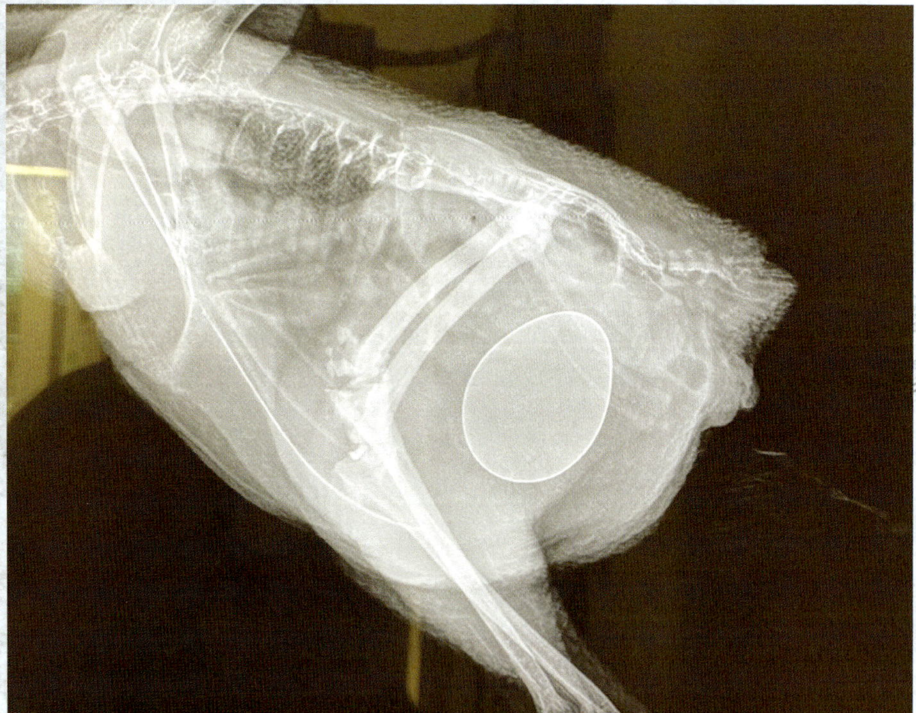

*X-ray of an atopic egg in a hen's abdomen.*

## Abnormal Eggs

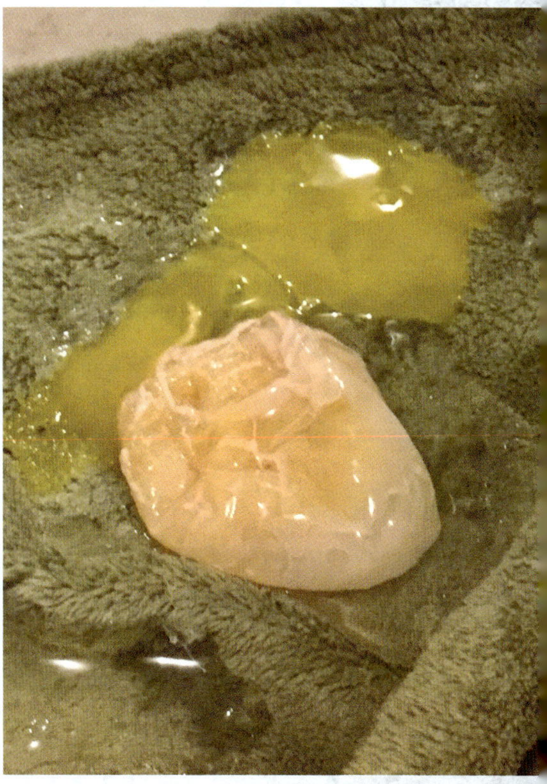

*Soft-shelled egg.*

Eggs can have an abnormal appearance for a variety of reasons. Eggs with soft, rubbery shells, very thin shells, or clusters of calcium spots on the surface can all be indications of calcium deficiency or problems absorbing calcium. Thin shells and eggs with a thickened band around the middle can be signs of heavy metal toxicity. Malformed shells including uneven spheres, wrinkled surfaces, and incomplete shells can be indicative of a variety of problems, such as vitamin and mineral deficiencies, reproductive system infections, mycoplasma infections, and reproductive cancer. Occasional abnormal eggs, often seen in new layers, senior hens, and molting hens coming back in to lay, aren't usually a cause of concern. A hen consistently laying abnormal eggs may have a serious problem and should be evaluated by your veterinarian.

## Vitamin and Mineral Deficiencies

Deficiencies generally present as abnormal eggs or a decrease in laying. They may also be passed along to any chicks produced by a hen with deficiencies. The most common deficiencies in hens with laying problems tend to be calcium and vitamin D.

## Infections and Salpingitis

Hens can develop infections in their ovary and their oviduct. The resulting inflammation, known as salpingitis, can cause difficulty laying eggs, abnormal eggs, lash eggs (these are not actual eggs, but rather clumps of infectious material), discharge from the cloaca, and general malaise.

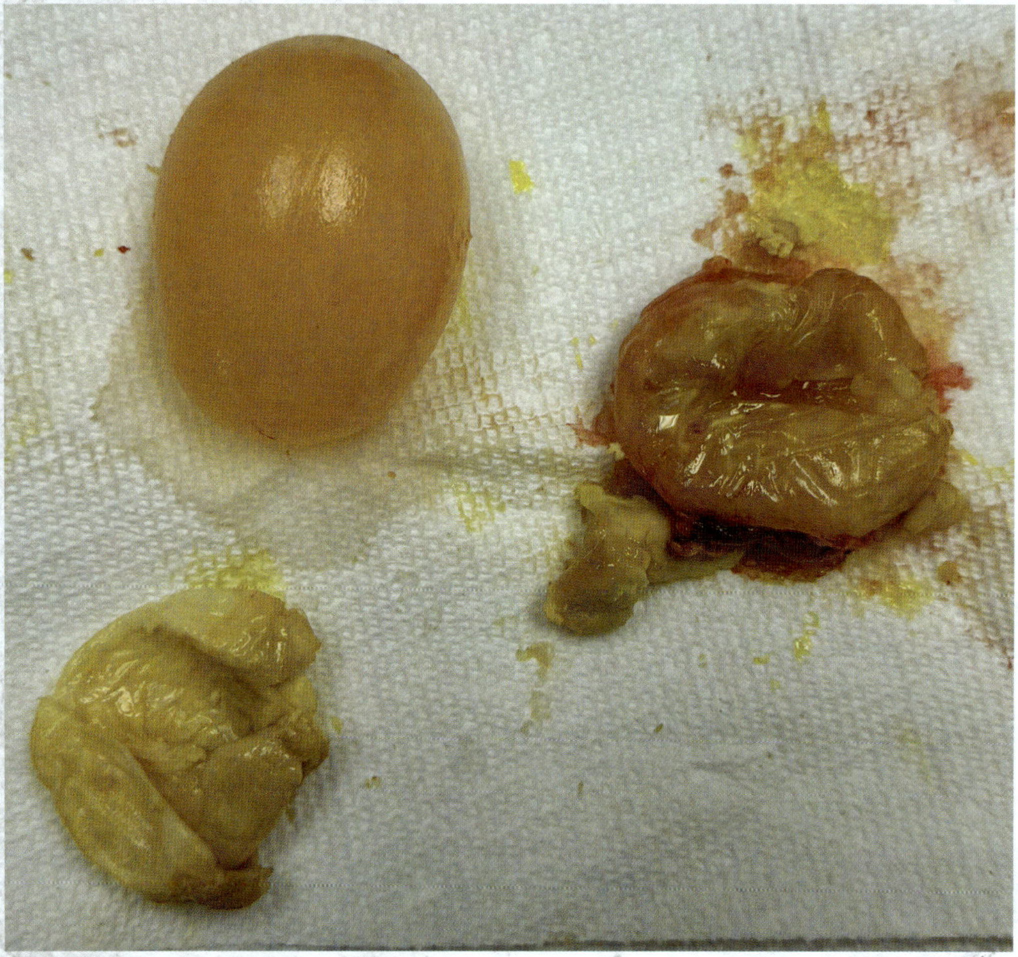

*Soft-shelled egg, lash egg, and infectious material.*

## Egg Yolk Peritonitis

Peritonitis occurs when egg yolk material enters the hen's abdomen because of a ruptured egg, an egg that hasn't fully developed, or an egg follicle that enters the hen's abdomen instead of descending into the oviduct. This causes inflammation and often a bacterial infection. Peritonitis generally requires long-term treatment with antibiotics and non-steroidal anti-inflammatories. This is a serious disorder that can lead to death if not treated.

## Vent Gleet

Vent gleet is a bacterial or fungal infection of the cloaca. Symptoms include thick white discharge with a foul odor, and a red and inflamed vent. (Do not mistake pasted-on urates for vent gleet. When you clean up the feathers, the vent will not be inflamed.) The vent may be red and inflamed. The hen may also stop laying eggs and look generally unwell. Treatment with antibiotics or antifungal medications is necessary. Your vet may also prescribe non-steroidal anti-inflammatories to keep the hen comfortable.

## Prolapse (Cloacal)

In prolapse, the lining of the vent or cloaca is pushed out and protrudes from the vent. The most common cause is straining to lay an egg. If the tissue is healthy, it should be cleaned and gently pushed back into place with a gloved hand. Inflammation can be treated by soaking in lukewarm water with Epsom salts or applying Preparation H to help shrink the tissue. Prescription non-steroidal anti-inflammatories may be required. Damaged tissue should be evaluated and treated by your veterinarian. Severe cases may require your veterinarian to place stitches to hold the tissue in place while it heals. Separate the hen so flock mates do not peck or injure the exposed tissue. Prolapse is often treated successfully, but it is a long game.

*Follicle tumors and other material removed from the abdomen of a hen with cancer.*

## Egg Binding

This is the term used when a hen is unable to lay an egg. There are a variety of causes, including calcium deficiency, an overly large egg, a very small egg, a soft-shelled egg, reproductive inflammation, infection, tumors, or cancer. Offer the hen calcium in the form of an eighth of a Tums tablet to help with contractions. Your veterinarian can also give the hen a fast-acting calcium injection and non-steroidal anti-inflammatories to treat pain and swelling. Keep your hen in a quiet, stress-free environment where she can lay her egg. Egg binding is a symptom of a larger problem. A hen who cannot lay in a day or two should be evaluated by your veterinarian.

### Tumors and Reproductive Cancer

Tumors and reproductive cancer can occur in any breed of hen, though they are most commonly found in hybrid layers. Your veterinarian can sometimes diagnose tumors and cancer via an internal vent exam. Some chicken keepers opt for exploratory surgery to determine the presence of cancer or tumors, but the hen must be in good-enough health to be anesthetized. Tumors can grow in every part of the reproductive tract, and some tumors can be removed during surgery. If the cancer isn't too far advanced, some hens can continue to live a good-quality life with maintenance medication.

### Best Practices for Reproductive Health

Laying hens should always have free-choice access to crushed oyster shell. Feed laying hens a good-quality layer's ration with adequate protein and calcium, as well as the recommended daily allowance of vitamins and minerals. Allow your hen's reproductive system to take its natural break when the days become shorter by not lighting your coop at night during the fall and winter. Ensure that your hens have time in the sunshine during laying season so that their bodies have enough vitamin D to allow calcium absorption. Try to keep your flock's environment as low-stress as possible; including an adequate number of nest boxes helps with this. Select and keep good-quality genetics in your flock.

## Temperature-Related Disorders

Heat stress, hypothermia, and frostbite are seasonal and generally caused by extreme temperatures in summer and winter. Heat stress and hypothermia can be fatal to your chickens. Frostbite is incredibly painful and can result in infection and/or the loss of toes, feet, combs, and wattles. Large hens, roosters, bantams and very small chickens, and very young and elderly birds are all at risk for temperature-related problems. Prevention is the key to avoiding these disorders.

## Heat Stress

Temperatures starting at 80 degrees Fahrenheit and upward can be enough to induce heat stress in birds that are not acclimated to it. It's very important to keep a close eye on your flock, but preparation is key. Sometimes a chicken that seems fine can be overcome by the heat, causing collapse and death very quickly.

*Poppy, a Barred Rock hen, cools off her feet in a kiddie pool.*

Studies have found that high temperatures can harm chickens in a variety of ways. They can cause your birds to lose electrolytes and/or become dehydrated. This can lead to dangerous pH changes in their bodies—too acidic or too alkaline—that can ultimately lead to organ failure. Heat can also cause oxidative stress in the chicken's body, in which there are too many free radical cells and not enough antioxidants to get rid of them.

Any chicken can succumb to heat stress or stroke, but especially big-bodied, fluffy birds and those with small combs and wattles. Bantams, Mediterranean birds, and birds with big combs and wattles fare better, but they can also be affected if temperatures are high enough.

Common signs of stress can include open-mouthed breathing, holding the wings away from the body, pale comb, listlessness and lying in the dirt for extended periods, and not eating and drinking. If you suspect heat stress, immediately move the bird to a cooler environment. Offer them water with electrolytes or watermelon if they're able to eat. Stand them in a container of cold water and place a cold compress on their comb.

## Hypothermia

The primary cause of hypothermia is temperatures so low that the chicken's body can't sustain itself. Organs shut down and the heart and respiration will stop. Danger of hypothermia increases when there are sustained temperatures below 25 degrees F. Elderly, special-needs, and small-bodied birds such as bantams are most at risk. You can move at-risk birds to a heated area or install radiant-panel heaters in your coop.

## Frostbite

Frostbite occurs when the cells in living tissue freeze and die from severe cold temperatures. There are degrees of frostbite: The least harmful stays on the skin surface, while the most severe can freeze so deeply that birds may lose combs and entire feet. The first areas to get frostbitten are usually combs and wattles. They're essentially exposed skin that is rich with blood vessels. Toes are also common areas for frostbite. Large chickens often have toes sticking out from under their feathers when roosting at night, and during the day, the toes are making contact with frigid ground. Roosters, with their big combs and wattles and large feet, are the most common frostbite victims.

*Small-bodied chickens appreciate supplemental heat in frigid cold. Little Man and Mona are living it up in front of the fireplace.*

It is a myth that chickens in a dry coop will not get frostbite or even die from hypothermia. While a damp coop encourages mold growth, and light frostbite from condensation settling on combs may happen, frostbite will still occur even if your coop is completely dry. Neither moist nor dry air stops frostbite. Because coops, climates, and chickens are all different, we suggest using a thermometer to monitor nighttime temperature. You can install a radiant heat plate with a thermostat cube to switch on the heater if the temperature in your coop drops below 20 degrees F (–6 degrees C).

*Kasper with mild frostbite on wattles and comb.*

Frostbitten areas will look pale or white at first. Over days and weeks the white spots will slowly turn black as the damaged skin dies and eventually falls off. During this time you will want to monitor the area closely for signs of infection. Minor cases of frostbite respond well to carefully applied aloe vera gel. In extreme cases you will want to take your chicken to your veterinarian for debridement and pain medication.

The most effective way to prevent frostbite is to make sure your chicken has a warm shelter, especially at night when temperatures are usually at their lowest. We highly recommend radiant panel heaters that can be used to keep temperatures at or around a safe 30 degrees Fahrenheit (–1.1 Celsius). Using a thermostat or ThermoCube to turn on the heater when necessary keeps your setup energy-efficient.

If you don't have power in your chicken run, the deep litter bedding method will add heat to a coop. Smaller, well-insulated plastic coops can have their interior temperature raised using hot water bottles or jugs of hot water placed inside. You also have the option of moving your chickens into a space such as a garage, basement, or barn, where you can set up temporary housing and supplemental heat. Petroleum jelly can stop windburn and the very lightest topical frostbite, but will not prevent deeper layers of tissue from freezing.

## Predator Attacks and Injuries

One of the biggest hurdles chicken keepers face is protecting their flock from predators. Everything wants to eat chicken. If you do experience an attack, work quickly to assess your bird's condition. Chickens are extremely resilient, and many survive wounds from predators. In extreme cases, your vet may need to debride and suture wounds, and can provide antibiotics and medication for pain management.

Injuries from predators range from punctures from teeth and claws to large open wounds and exposed tissue. Your bird may also be in shock. Get them to a warm, safe place, such as your chicken hospital. You can give them Poultry Nutri-Drench or electrolyte water and then call your veterinarian to schedule an appointment. Assess the wounds and clean with chlorhexidine and water. Seal the wound with honey—right from your cabinet is fine—to keep the wound moist and fight infection until your vet can see the chicken. Monitor them closely; you may need to offer liquid food if they are not eating and drinking enough.

Be vigilant about checking your coop and run for predator access points, and make timely repairs. Supervised free-range time can help cut down on the risk of predator attacks. Some chicken keepers and farmers with large flocks choose to use trained livestock guardian dogs to keep their flocks safe.

## Marek's Disease

Marek's disease is a virus that primarily affects chickens. It is a herpesvirus that passes from bird to bird through feather and skin dander. Once a chicken is infected, it is a carrier for life. There is a vaccination available for one-day-old chicks that is extremely effective, and we encourage all new chicken keepers to purchase vaccinated chicks.

Young chickens, especially those between six weeks and eight months old, are most at risk of developing the disease. If you have Marek's in your flock, your birds will be contagious for many years. The virus survives in dust and dander for up to three years.

The disease is named after József Marek, the Hungarian veterinarian who first described it in 1907. There are several different strains of the virus, and they manifest in chickens in many different ways. Neurological problems such as lameness and leg paralysis is one of the more common forms. Other forms include eye abnormalities, skin tumors and other abnormalities, and cancerous tumors on organs. There is no treatment for Marek's disease, and there is no cure. Palliative and supportive care are recommended to keep a bird with symptoms comfortable.

Besides vaccination, biosecurity is one of the best defenses against Marek's infecting your flock. Always quarantine new birds, and do not purchase chickens from auctions or from anyone who does not vaccinate for the disease. Practice good shoe hygiene: Keep one pair of shoes that you only wear with your flock and nowhere else. Don't let others who keep chickens wear their coop shoes with your birds. Biosecurity may seem a bit over the top for a backyard chicken keeper, but it helps keep Marek's, avian flu, and a lot of other serious diseases away from your birds.

## Finding a Veterinarian and When to Call Them

We recommend finding an avian veterinarian as soon as you decide that you want to have chickens. Sometimes you have to do a lot of looking and calling around to find someone willing to treat poultry. Many veterinarians have not been trained to treat pet poultry, only commercial operations. Some farm and large-animal vets are happy to see chickens, but avian veterinarians in small animal hospitals are also an excellent option. Once you find a vet, schedule a well visit for at least one of your birds so that you're officially a client. That way, you'll be able to get an appointment if you have a sick bird or some other emergency.

We are not sure why so many members of the backyard chicken community think you cannot take a chicken to see a veterinarian. All living things deserve medical care, and your beloved chickens are just as important as any other animal. You may have to drive a bit to get to a good vet, and you may need to advocate for your birds with a vet who doesn't have a lot of poultry experience, but it will be worth it to have access to treatment and medications when your birds are ill or injured.

You'll want to call your veterinarian for birds that won't eat or drink for more than a day; that lie down and don't want to walk; that have abnormal crops or complicated bumblefoot; that experience predator attacks; that have breathing issues, reproductive problems, or any ailment that needs prescription medication. If you are hatching chicks, many vets will administer the Marek's vaccine to day-old chicks if you purchase the vaccine.

## Special-Needs and Senior Chickens

*Bubbles, a senior Buff Orpington hen.*

If you talk to people who keep pet chickens, you'll quickly discover that it's not at all unusual for Chicken Ladies and Gents to have special-needs and senior chickens (those over five or six years old) in their flock. With more information available about good chicken husbandry and access to medical care, our chickens are living longer, healthier lives. We have more resources and information today that help us care for special-needs birds. Conventional wisdom tells us to cull birds with issues, but don't believe the hype. You can successfully keep special-needs chickens and give them excellent quality of life if that is what you want to do.

Chickens can be born with genetic problems and birth defects, but sometimes they end up as special-needs chickens because of illness or injury. We'll mention some of the most common issues, and then share our best tips for taking care of our precious senior hens and roosters and our special-needs birds.

### Crossbeak

Crossbeak is a disorder affecting the alignment of a chicken's beak. The bottom beak, which is attached to the jawbone, can be misaligned. It can range from mild—where the beak still points forward but the top and bottom do not match up—to severe misalignment, where the bottom beak points sideways. Chicks can hatch with a genetically induced crossbeak, but they can also develop crossbeak from injuries.

In the mildest forms, the chicken can still eat and drink with little to no problem. In the most severe form, the chicken will need to be fed via syringe, or "torpedo feeding," where a mash is shaped into bullets and placed into the bird's mouth. In our experience, mash is a good idea for all crossbeaks. Not only does it make torpedo feeding possible, it also helps birds with a less severe crossbeak scoop up the food a lot more easily. Monitor your crossbeak birds closely to make sure they are a healthy weight. Chickens with severe crossbeaks may need help preening their feathers.

## Neurological Issues

Chickens can be born with neurological issues, but problems can also develop later on from illness and injury. Neurological issues can manifest in several ways, including

*Clover manages very well with her neurological issues.*

head tilts, balance problems, and sometimes an odd gait as the chickens walk or run. As long as they are thriving and not being picked on, these birds can safely stay with their flock. Many of them prefer sleeping in bedding instead of roosting. Some neurological problems can result in paralysis. These birds generally need to be housed and kept by themselves. We have had great success using chicken strollers and chicken wheelchairs to help them get around.

## Sight-Limited Chickens

Chickens can be blind at hatch, or they can lose sight in one or both eyes from injury or illness, including ocular Marek's, conjunctivitis, pecking injuries, or puncture wounds from materials in their environment. While there is no treatment for ocular Marek's, conjunctivitis and other infections and injuries should be seen by a vet, and can be treated with Terramycin or prescription antibiotics.

Some blind chickens are lucky enough to have a buddy that helps them navigate everyday life. Other birds may need to live outside of the flock and with special accommodations to allow them to easily find their food and water. It will improve their quality of life and lower their stress enormously to have a friend live with them. Consult with your veterinarian to ensure that your chicken's eyes are receiving any ongoing treatment that may be necessary.

Chickens can lead fairly normal lives with the loss of one eye. This is almost always the result of injury, and you should consult with your vet to make sure the injured eye receives any aftercare necessary to keep it healthy and free of infection. In some cases, the injured eye will be removed.

## Mobility Problems

Senior chickens may develop arthritis and find getting around a bit harder. Sometimes these birds will even begin waiting for you to lift them on and off of the roosts at bedtime and first thing in the morning. Keep an eye on them to make sure they're eating and drinking enough and that the rest of the flock are not bullying them.

*Anastasia, a hen with limited mobility, relaxes on a chair in the run.*

## Pain Management

In cases where your birds may have ongoing pain due to an old injury, arthritis, neurological problems, etc., your veterinarian can prescribe pain medication and non-steroidal anti-inflammatories. Arnica gel or ointment, long used by humans to treat pain and bruising, may also offer some relief when applied topically to ouchy spots on your chickens. Consult with your veterinarian for proper dosage.

## Supplemental Nutrition

Both special-needs and senior chickens may need supplemental nutrition to keep them as healthy as possible. This can be as simple as making time for them to eat and drink extra on their own or syringe-feeding them with a powder-to-liquid food such as EmerAid Intensive Care Omnivore or Oxbow Critical Care.

## Dust-Bathing and Feather Care

Birds with mobility issues, neurological issues, crossbeak, and senior birds with arthritis may require help with skin and feather maintenance. Often, they can't get to those hard-to-reach places, and some of them aren't even able to spread dust on themselves. You can give them periodic dust baths, and you may also need to help them remove the keratin sheath on newly emerged feathers. You can gently and carefully pick or scrape the sheath off of the feathers.

As we mentioned in chapter 3, chickens rely on dust-bathing to clear excess oils from their feathers and smother the occasional mite of chicken louse that tries to hitch a ride. It is easy enough to help your birds do this. Using a basin, box, or some other wide and shallow receptacle, place a few inches of dust-bathing material inside. You can just get some dirt, or you can make them a special mix; some dirt, sand, and wood ash make a very nice blend. You can also use or add some of the dust-bathing blends that are commercially available. Many of them contain essential oils that help repel creepy crawlies and maintain feather health.

Place your chicken in the basin and gently sift the mixture over the top, lifting his or her wings and trying to work this dirt bath down to the skin. Be careful to avoid the eyes, and try to avoid making dust clouds that both of you may accidentally inhale. When bath time is over, lift your bird out and take him or her someplace (we just find a patch of lawn) where you can gently shake and brush off the dust. You could also use a hairdryer on low to help blow out the dust.

## Coop and Run Modifications

Lower roosts and longer, more gradual ramps can make a difference to birds that have balance or mobility problems. Make sure there is plenty of food and water that your senior and special-needs birds can reach. In some cases, you can partition off a section of run that is just for your special girls and boys. You may be able to have them sleep in the coop, possibly in the bedding, and then tuck them into their own space during the day.

## End-of-Life Care

It would be wonderful if our feathered friends could stay with us forever. Still, we can do our best to make the end of their lives as comfortable as possible. You should discuss options with your veterinarian. There are options for palliative care, and humane euthanasia is also available.

*Placing a shallower wooden ramp over metal steps makes life easier for senior chickens.*

*Dr. Rebecca Gounaris with one of her earliest poultry patients, an Easter Egger hen.*

CHAPTER 10

# Remarkable Chicken Ladies

Throughout time, there have probably been millions of women who kept chickens. It would be great if we knew more about them and could share that information with you. Unfortunately, chicken keeping was considered women's business, and not important enough to write about until the 1800s. Though we can't bring you much in the way of early Chicken Lady history, we can share some more recent stories with you. The following eight women have made significant contributions to the modern chicken-keeping world, and they're all amazing in their own right! Their hard work has paved the way for all of us to become successful Chicken Ladies.

## Queen Victoria, the Original Chicken Lady

◀ *Young Queen Victoria.* PHOTO CREDIT: FRANZ XAVER WINTERHALTER, QUEEN VICTORIA (1819–1901), ROYAL COLLECTION, WIKIMEDIA COMMONS

Early in this book we mentioned the fact that throughout most of our recorded history, women were in charge of the poultry. Some archaeological evidence points to chickens that were kept as pets, but there isn't much documentation that illustrates what that looked like, practically speaking. This changed when Queen Victoria ascended the British throne.

Born in 1819 at Kensington Palace in London, Alexandrina Victoria was an animal lover from the time she was young. She kept pet dogs of all kinds, including terriers, spaniels, and greyhounds. One of her favorites was Dash, a Cavalier King Charles spaniel that she kept in her younger years. As a young woman she also developed a love for birds, and kept a number of parrots and tropical birds.

Victoria took the throne in 1837 and married her beloved husband, Prince Albert of Saxe-Coburg and Gotha, in 1840. Together they filled the grounds of Windsor Castle with even more animals. In 1842, Victoria and Albert had the royal aviary on the grounds of Windsor remodeled. And just in time, because later in 1842 (some sources say 1843), Colonel Edward Belcher presented Her Majesty with a gift of five Cochin-China hens and two cockerels. All reports claim that Victoria was enthralled with her birds, visiting them as often as possible so she could sit and watch them.

At the time, the young queen and her prince consort were trendsetters. They brought everything from Christmas trees to white wedding dresses to heritage-breed chickens into fashion. The queen received more gifts of fancy poultry, and before long, chickens were the "in thing." Hen Fever, a purebred chicken bubble that affected both the United Kingdom and United States, inflated prices until people were paying the equivalent of thousands of dollars in today's money for their birds. Eventually Hen Fever subsided, but Queen Victoria and many of her devoted followers remained Chicken Ladies and Gents. Fancy chicken breeds were here to stay.

## Miss A. C. Croad, Champion of the Langshan

In 1872, a shipment of glossy black chickens arrived in England for Major F. T. Croad. Named for Langshan, the area of China from which they came, the chickens had been discovered by a relative of Major Croad who was doing work for the British government in the interior of China.

But the birds didn't remain with Major Croad. These large, glossy black chickens owe much of their preservation

*We could find no photos of Miss Croad; instead we pay tribute with a vintage painting of her favorite chicken breed, the Langshan.* ▶
PHOTO CREDIT: PRIZE AND GAME CHICKENS (N20), GAST LITHOGRAPH & ENGRAVING COMPANY, JEFFERSON R. BURDICK COLLECTION / GIFT OF JEFFERSON R. BURDICK

and popularity to the work of his niece, Miss A. C. Croad. In the United Kingdom, the original form of the breed is known as the Croad Langshan. This tribute is not for Major Croad; it is for his niece and her tireless work with this beautiful breed.

We have searched high and low for the first and middle names of Miss Croad, but we have been unable to find them. Use of initials was very common during the 1800s, so she remains Miss A. C. Croad. In much of her writing and correspondence, she did not use a feminine title, even answering to Mr. Croad. In some of the letters she exchanged with poultry historian Lewis Wright, she requested that he keep her identity a secret. Word of her gender did eventually get out, but we don't know if Wright was the one who let the cat out of the bag.

Many of the poultry fanciers who saw the first Langshans decided that they were not a distinct breed but simply an inferior strain of Cochin. Miss Croad disagreed strongly and fought for years to have them accepted as a breed with its own merits. In 1877 she wrote a book, *The Langshan Fowls: Their History and Their Merits.* Thanks to Miss Croad, Croad Langshans went on to become a respected breed, both in the show ring and on the farm. Several modern breeds, including the Black Orpington, Marans, and Barnevelders, have Langshans as one of their foundation breeds.

## Mrs. Nettie Metcalf, Creator of the Buckeye

◀ *Nettie Metcalf, the first woman to develop an American breed chicken.* PHOTO CREDIT: AMERICAN BUCKEYE CLUB PUBLICATION, CA. 1910s

Nettie Williams Metcalf, born in Ohio in 1859, wanted to breed a productive red bird that did well in Ohio winters. She married Frank Metcalf in 1879, and like most new wives of the time, she got started right away with her first chickens, a flock of Brown Leghorns. The Leghorns did not work for her, nor did the Black Langshans and Barred Rocks that succeeded them. So, she began an experimental breeding program, crossing

Barred Rocks and Buff Cochins. The following year she added in game birds, and the offspring turned out to be red chickens.

Nettie Metcalf wrote that her breeding efforts were often laughed at, but this did not deter her from her mission. She continued breeding through the 1890s, until she produced chickens that met all of her goals. They were a rich, dark mahogany red color, large and docile, but active, dual-purpose birds. Their pea combs and tight feathering helped them deal with cold weather, and the hens were very respectable layers of anywhere from 150 to 200-plus eggs per year. They would also go broody and hatch chicks, an important feature in the late 1800s and early 1900s.

The Buckeye chicken was accepted into the American Poultry Association's *Standard of Perfection* in 1905. Nettie Metcalf remains the only American woman to have created a chicken breed officially recognized by the APA.

## Jane Howorth, Founder and Director of the British Hen Welfare Trust

As a teenager, Jane Howorth saw the realities of industrial egg farming and knew that she had to do something to help the chickens. In 1995, Jane went to a local egg farmer and asked if she could have some of the "spent" laying hens that were scheduled for slaughter. She intended to take a dozen, but she ended up with thirty-six of them.

In 2003, Jane placed an ad in a local paper, offering people the chance to adopt ex–battery cage hens. These hens are kept their entire lives in "battery" cages so small that they can barely turn around. Jane's idea was so successful that she started organizing adoption events. She also decided to work with the egg industry instead of pointing fingers and placing blame. This allowed her to help many more chickens, giving them a chance to experience sunshine, grass, and love.

Jane's rescue efforts were the beginnings of the British Hen Welfare Trust. She and her amazing team of volunteers have worked to successfully end battery cage farming in the United Kingdom. In 2024, the millionth hen rescued by the British Hen Welfare Trust, Henrietta, was adopted by King Charles. Henrietta and thirty of her friends went to live at Highgrove House in Gloucestershire.

Jane is a very special human being, and one of the most inspirational women we have ever known. She was awarded the MBE, Member of the British Empire, for outstanding community service in 2016. The British Hen Welfare Trust has

*Jane with happy ex-battery hen.* PHOTO CREDIT: SMGDC, LTD.

significantly changed the way that people view chickens, especially commercial hybrid layers. The organization is also dedicated to educating and assisting chicken keepers with hen health care.

## Fiona Osborne, Breeder of the English Orpington

Fiona Osborne has been breeding English Buff Orpingtons for more than a decade. She and her husband, Hugh, are the duo behind *English Country Life*, and are known for their inspiring and instructional homesteading videos, recorded at their smallholding in the beautiful English countryside. They produce much of their own food and other products on one acre of land.

Fiona has devoted much of her time to teaching chicken keepers how to safely and effectively use broody hens to hatch eggs. She provides ongoing education on how to set up a breeding program, select parent stock, and follow best practices for raising your own heritage-breed chickens. Fiona and Hugh's Buff Orpington flock are bred to standard and are some of the most beautiful chickens you will ever see.

*Fiona with her beloved Orpingtons.*

## Dr. Rebecca Gounaris, Poultry Veterinarian

Dr. Rebecca Gounaris is a third-generation veterinarian, following in the footsteps of her father and grandfather. A graduate of the University of Delaware and Virginia Maryland College of Veterinary Medicine, Dr. Rebecca spent several years serving as a volunteer and intern with Tri-State Bird Rescue in Delaware. After graduation, she came home to work at the family practice, Pleasantville Animal Hospital in Fallston, Maryland. Her skill as an avian veterinarian attracted local chicken keepers who needed medical care for their flock. In fact, her first-ever patient at the hospital was a duckling.

*Dr. Rebecca with Cornelia, a Legbar hen, and a recent award.*

Chickens are Dr. Rebecca's favorite patients because of their resilience; she encourages their people to give them a chance if they're ill or injured. She believes that chickens are smarter and stronger than people think. After seeing chicken patients for three years, Dr. Rebecca started a flock of her own. Her favorite type of chicken is the Easter Egger—they're adorable, generally healthy, and tend to be long-lived.

We are privileged to have Dr. Rebecca treat our birds. She is dedicated to caring for pet chickens and other poultry, and continues to learn new practices and medical treatments. She would like to see more practices become open to treating backyard flocks. Veterinarians like Dr. Rebecca make the world a better place, one chicken at a time.

## Jeanette Beringer, Senior Program Manager of the Livestock Conservancy and Conservator of the Crevecoeur

Jeanette Beringer has served as a senior project manager with the Livestock Conservancy, an American nonprofit organization, for over twenty years. She is dedicated to the work of preserving rare and heritage breeds, including poultry. Before working with the Conservancy, Jeanette was the first female head zookeeper at the Roger Williams Park Zoo in Providence, Rhode Island.

Jeanette has worked with various livestock breeds over the years, including the Buckeye chicken, the Nankin bantam, and the Crevecoeur chicken. She chose to document her work with the Crevecoeur on a Facebook page, sharing history, tips, and information about breeding these birds that could not be found elsewhere. A dramatic-looking old French breed, the Crevecoeur is an all-black chicken with a large crest on its head. When Jeanette began her program, this breed was critically endangered. As she prepares to wrap up her breeding program, the most recent poultry census has moved the Crevecoeur past the Threatened category and into the Watch category.

Jeanette is well known for her commitment to heritage breeds and the people who work with them. She is an excellent spokeswoman for critically endangered livestock, speaking at conferences and events and educating people on the joys of keeping chickens.

*Jeanette with a Sicilian Buttercup Cockerel.*

## Lisa Steele, the Original Chicken Blogger

Lisa Steele has filled many roles in her time, including working as a Wall Street accountant, a bookstore owner, a chicken blogger, and a cookbook author. In 2009, Lisa found herself married and living in Virginia, and decided to get back to her roots as a fifth-generation chicken keeper. She bought chickens and started photographing and writing about them, and in 2011, her blog, *Fresh Eggs Daily*, was born. Lisa wrote about keeping chickens in your backyard with a focus on preventive care for pet chickens—something that no one else was doing at the time.

*Fresh Eggs Daily* went beyond basic care, though. Lisa wrote about using fresh eggs to cook and bake, making your coop fancy with nest-box curtains, and hanging streamers of bunting in the runs. The natural outgrowth of all this blogging was a book. *Fresh Eggs Daily* was published in 2013, and was followed by a book about raising ducks, a book about gardening with chickens, and three other books on chicken keeping. In 2022, *The Fresh Eggs Daily Cookbook*, Lisa's first cookbook, was published. Lisa hosted a TV show on PBS for two years, *Welcome to My Farm*, and she has authored several cookbooks. Her blog and her books have inspired countless people to start their own backyard flocks.

*Lisa with a lavender hen.*

*Chicken Lady besties.*

CHAPTER 11

# THE CHICKEN LADY LIFE

Anyone can be a Chicken Lady. All you need is a chicken, or twenty. Or thirty. You'll spend a lot of your time thinking about chickens, and you'll start seeing chickens everywhere. This includes (but is not limited to) plates and dishes, Christmas decor (three French hens!) and home decor, in movies and on TV, clothing, jewelry, and even lawn decorations. You may also find yourself the recipient of chicken gifts. We've been gifted everything from a chicken cutting board to a chicken pool float.

That's the fun part of Chicken Lady stuff. There's also the very real responsibility that comes with chickening. Your schedule may change, and you may find yourself jealously guarding your flock time. Don't be surprised if you suddenly find yourself telling people, "I can't, that's chicken bedtime." Getting chickens is making a commitment to their care and their emotional well-being. If you've never had animals before, you may be a bit surprised by the wonderful ways that they can take over your life.

## Chicken Lady Network

The Chicken Lady network is a fast-growing group of people—both online and in person—who are in love with their chickens. Finding like-minded people can really help if you need someone to commiserate with you on a loss, or to celebrate those first eggs with you. People keep chickens for all sorts of reasons, and in all different ways. Being able to share your thoughts with someone whose chickens are just as beloved as yours is validating and a source of joy and comfort. You may also become very close friends and maybe even start thrifting together for everything chicken-themed!

*Next-level Chicken Lady life.*

*Thrifting hard at a Goodwill in Ohio, 2022.*

## Collecting Vintage

Enough chitchat. Time for the good stuff: vintage chicken decor. We both love history and antiques, and we love the fact that chicken decor was so common for so long. If you're already into buying secondhand, then you'll understand the draw. If you're just venturing into thrifting or collecting, buckle up. It's fairly addictive, but you always have the power to curate your collection. You may decide that you only want to collect egg cups or table linens. You can limit your collection to one shelf, or to one glorious chicken-filled cabinet. (Fair warning: That would be hard to do.)

Over the years we have become very selective about the items we choose to buy. Don't worry, there's still plenty of good chicken stuff out there for you! We're going to cover some of our favorite things to collect, but there are lots of other things that we can't cover in detail. That includes soup tureens, gravy boats, teapots, serving plates and platters, glassware and barware, napkin rings, mugs, vases, and pitchers.

## Deviled Egg Plates

We have *a lot* of deviled egg plates, and they make a great place to display your fresh eggs. They're easy to find at thrift shops and yard sales, generally cost less than $10, and they're also relatively easy to store. The common egg plates that you'll find are usually clear glass or milk glass. The milk-glass plates are often from the 1950s to 1970s, and make a pretty addition to a spring table. More ornate dishes may include painted chickens, or even figural chickens in the form of dip bowls and salt and pepper shakers.

*A few of our favorite egg plates.*

*A selection of egg cups in our collections.*

## Egg Cups

Egg cups are a lot of fun to collect. They don't take up tons of space, and you can use them for tablescapes and displays. They're not always easy to find, which amps up the thrill of the hunt, and some of them can be quite expensive. But we've put together a very nice collection over the years, and you can too. Some of our favorites: fine china blue willow cups, chick-shaped figural cups, crystal cups, milk-glass and jadeite cups, and hand-painted Pennsylvania German cups.

## Egg Scales

This is a collection you'll probably build rather slowly. There are antique scales out there, and plenty of reproductions, too. Some of them still work, and it's both helpful and a lot of fun to have them in your kitchen. Kids *love* to weigh eggs on the scale.

*Our favorite cookie jars.*

## Cookie Jars

Some of our favorite kitchen decor are our cookie jars, and we do keep cookies in many of them. Obviously, they take up more space than a lot of other collectibles, but they can be useful, and they are generally adorable. We've been known to use them as table centerpieces as well. Add some fresh flowers, and you're on your way to a cute and cozy tablescape.

## Hen-on-Nest Dishes

Many of us remember first seeing these quaint dishes at our grandmother's house. Usually made of glass, hen-on-nest dishes have been produced in the United States since the 1870s. Challinor made some of the earliest dishes, but Indiana Glass Company, Westmoreland, and Hazel Atlas are great producers to collect as well. They come in various sizes and lots of different colors, including milk-glass white and blue cobalt glass. Some people use them as candy dishes, but they're also great for storing trinkets and jewelry, kitchen condiments, or even eggs, in the largest dishes. If you're lucky, you may find a rooster-on-a-nest dish.

*Jadeite hen on nest.*

*Mid-century modern salt and pepper shakers.*

## Salt and Pepper Shakers

Salt and pepper shakers win our vote as one of the most collectible chicken items. They're small and usually easy to store and display, and they are generally on the lower end of the price scale. One of our absolute favorite things to do with salt and pepper shakers is to use them as cake toppers. They also look great on a windowsill in your kitchen. While most of them are ceramic or porcelain, you can also find them in pewter, jadeite, milk glass, wood, and chalkware.

## Statues and Figurines

These come in many, many shapes and sizes. From very large statues, mostly of roosters, to the tiniest hens, chicken figurines often show up at our favorite thrift shops. And we buy them. They're super fun to display anywhere you have a bit of extra space. The top of the refrigerator, tucked next to the legs of a coffee table, the fireplace mantel, on top of a stack of books on a shelf, in a country hutch, in an entryway, or even right in the middle of your dinner table. We have many in our recording studio. Some of the best are mid-century modern, but we don't discriminate when it comes to chicken statues and figurines. Some of our favorite makers include Lefton, Royal Haeger, Fitz and Floyd, Napco, Otagiri, California Pottery, and Royal Copley. While most figures are ceramic or porcelain, you might also find them made from carved wood, blown or molded glass, metal, cast iron, brass, and resin. You may find some for a few dollars, while others cost hundreds. Buy what you love.

*Some of our chicken statues.*

## Hosting and Entertaining

Entertaining is not a required part of being a Chicken Lady, but you may find yourself enjoying it if you make it chicken-themed. You can show off your collected treasures and use some of those delicious eggs in your recipes. (Of course, there's nothing wrong with putting some flowers and a chicken statue on your table to enjoy yourself. This is your Chicken Lady journey—do it your way.)

## Tablescapes and Settings

We absolutely love creating tablescapes. If there were a Chicken Lady tablescape competition, we'd be all in. We train so hard that we'd be ready for the Chicken Lady Tablescape Olympics. There are so many ways that you can incorporate chickens and eggs into a table setting. The most obvious and the easiest way to do this is to use chicken dishes or china. The options run from very casual plates to fancy full sets. There are also plenty of vintage sets of beautiful glassware featuring chickens that you can use on your table. Just a note of caution: Some antique sets may have unsafe glaze and should not be used to serve food or beverages.

*One of Chrisie's holiday tablescapes.*

*One of Holly's Thanksgiving tablescapes.*

If you don't want to use chicken plates or drinkware, you could use chicken-themed serving ware instead. Large platters and serving trays as well as gravy boats and soup tureens are good options. There are some large tureens that are figural, in the shape of a chicken. They tend to be expensive but make a beautiful focal point on your table. You can also create a centerpiece using chicken statues or figurines. Add some flowers or greenery and/or eggs, and you're on your way to a beautifully layered table.

Chicken-themed linens are another fabulous alternative. Tablecloths, napkins, placemats, and table runners are all possibilities. We each have a couple sets of adorable chicken-centric napkin rings. Holly also has a set of small white china roosters that are intended to be place-card holders and look adorable sitting at table settings. They work well for this, but you can also use them for other purposes. They can hold menus and small signs labeling dishes in a buffet. You can also repurpose salt and pepper shakers as place-card holders.

If you want to try tablescaping but aren't sure where to start, pick something to be your focal point. It could be a color found in your favorite chicken cookie jar, a print on your napkins, or, if you're looking for a challenge, try a specific breed. Gather items that "go with" your focal point. You can use a limited color palette, or go bright and wild. You can lean into a country or rustic look, or keep things formal and elegant. Layer in items that support your overall theme, and don't forget to add texture with linens, plants and flowers, and even paper and vintage ephemera. And don't be afraid to edit your creation. You have full control; take some things off, add more, move some things around. There is no wrong way to do this. The only goal is a table setting that you love.

Holidays are a great time to pull out the poultry decorations. Easter is perfect for chicks and eggs and maybe a duckling or two. Thanksgiving is turkey time. If you have any vintage turkeys, work them into your tablescape! We feel that geese are very appropriate for Christmas, but don't forget those three French hens.

## Outdoor Garden Parties

We *love* a good garden party. Whether it's afternoon tea, cake and coffee, Chicken Ladies who lunch, happy hour by the coop, or even a Chicken Lady wedding, we're there! We love to bring our chicken-y treasures outside and put together a festive setup. Break out those beautiful teacups featuring chickens and show off your vintage tea- and coffeepots with hens on the lid or roosters painted on the sides. We especially love a buffet table where we can layer vintage sets of salt and pepper shakers, hen-on-nest dishes, and small figurines. Candlesticks/holders, butter dishes, sugar bowls, creamers, pitchers, and jelly pots all make great additions, and don't forget those chicken-printed cushions and pillows for your fancy chairs and benches. Egg cups look adorable holding eggs, or you can repurpose the larger cups as vases. And make sure you feature plenty of egg dishes in your menu.

*Buffet table with chickens at our first Chicken Lady garden party, 2021.*

*Chicken Ladies at the garden party.*

## HOLLY'S WEDDING

Holly and Pete got married on October 19, 2019, in a chicken-themed barn wedding. The place cards had chickens stamped on them, the table numbers were made with vintage chicken breed images, and one of the cakes even featured chickens. They served three different types of lasagne—cheese, seafood, and butternut squash—all made using delicious, fresh eggs from their chickens. Holly also used their eggs to make the wedding cakes: a chocolate cake filled with raspberry cream, and a spiced pear cake with maple frosting.

*Chickens on the wedding cake and chickens on the tables.*

## Decorating Coops and Runs

Like all of our Chicken Lady activities, decorating your coops and runs is optional. Chrisie is fantastic at pulling out all the stops with bunting, colorful seating, and seasonal wreaths. Holly actually uses very minimal decoration, preferring to keep the focus on the natural backdrop of her farm. There are plenty of creative options if you want to fancy up your coop and run—wallpaper, chandeliers, rustic lighting, curtains, fairy lights, benches and seating, contact paper, wreaths and greenery, wind chimes,

*The DiCarlo coop all decked out for the Fourth of July.*

planters, flags, decorative solar lights, decorative garden stakes, and all kinds of signs. Your local farm and garden supply shops have plenty of fun Chicken Lady accessories available. You can even put a giant metal rooster in your yard, or set out a giant rooster skeleton during spooky season. Or how about a giant hen dressed as Mrs. Claus? As long as your flock is safe, go ahead and have fun with the decor!

## Morning Coffee with Your Flock

For us, there is nothing more relaxing than releasing the chickens and sitting down with a hot cup of coffee to enjoy our flock. This is the essence of the Chicken Lady life: carving out time to be with your feathered babies. Our best advice is to try to do this as much as possible. You may end up with some hen snuggle time, or your flock may be busy foraging for breakfast. Either way, you get to soak it all in, and your birds will love knowing that you are there with them.

# Delicious Egg Recipes

Chickens are the original friends with benefits—gorgeous, fluffy, huggable, and they give you all of the eggs! We spent chapter 2 on a deep dive into all things eggy, but now we're going to share some of our favorite egg recipes with you. There are so many ways to use them, and not just for breakfast (though we do love a Belgian waffle or some delicious French toast). Eggs are a foundation of virtually every cuisine in the world, and we have great fun experimenting with and developing new dishes or variations of old favorites. A couple of our egg-related mantras: Everything tastes better with a fried egg on top, and eggs and pasta are a match made in heaven.

*A delicious, homey egg recipe.*

# Judy's Christmas Egg Bake—A Savory Strata

*Yields 8–12 servings*

This recipe comes from Chrisie's mother-in-law, Judy. It remains a tradition in the DiCarlo house to serve this dish on Christmas morning. It is essentially a very delicious savory bread pudding. You can make it with any type of sausage, including meatless or veggie sausage, or leave the meat out entirely and add 3 cups of your favorite cooked veggies. It's also very easy to convert this dish to gluten- and dairy-free.

The egg bake needs time for the bread to soak up all of the eggy goodness, so plan ahead by making this the day before and refrigerating overnight.

## Ingredients

10–12 eggs, depending on size

2 cups milk (or unsweetened soy, oat, or almond milk for dairy-free)

1 teaspoon salt

1 teaspoon pepper

1 teaspoon dried mustard

1 pound sausage, cooked and drained (or 3 cups of cooked vegetables or any combination of sausage/veg)

2–4 slices of bread, cubed (or 4–8 slices gluten-free bread, depending on loaf size)

1–3 cups shredded cheese of your choice (or dairy-free cheese)

## Instructions

1. In a large bowl, beat together eggs and milk. Add salt, pepper, and dried mustard.
2. In a large 9×13 baking dish layer the bread cubes, sausage/veg, and cheese. Pour the egg mixture over the top of the layers, cover, and refrigerate overnight.
3. In the morning, bake at 350 degrees for 45–50 minutes. Let the egg bake rest for about 10 minutes, then serve.

*Custard topped with whipped cream.*

# Baked Egg Custard

*Yields 6 servings*

Classic egg custard has a delightful taste and texture that isn't easily described. It is a wonderfully old-fashioned dessert that is delicious, especially if it's made with rich, fresh eggs from our own flock. We bake this in custard cups or ramekins, but you can also pour it into your favorite crust and have it as a pie.

## Ingredients

- 2 cups milk (or oat or another plant milk)
- 4 eggs
- ½ cup sugar
- ¼ teaspoon salt
- ½ teaspoon vanilla extract or vanilla paste
- Ground nutmeg

## Instructions

1. Preheat oven to 350° F.
2. Start by scalding the milk in a small saucepan. Heat the milk over medium heat until it is steaming and you see small bubbles around the edges of the pan. Don't let it come to a boil. Remove from heat and set aside.
3. While the milk is heating, in a medium bowl whisk together the eggs, sugar, salt, and vanilla.
4. Temper the eggs. Whisking constantly, pour a small amount of the heated milk into the egg mixture. You want to slowly bring up the temperature of the eggs without making them cook. Gradually add more of the milk while whisking.
5. Once you've mixed in all of the milk, pour the mixture into custard cups (ramekins) or your pie shell. Sprinkle the top with ground nutmeg.
6. Place custard cups in a baking pan and then place in the oven. Carefully pour about an inch of hot water around the cups and close the oven door. The hot water allows the custard to cook slowly and evenly, developing a silky, custardy texture.
7. Bake for 40–45 minutes, or until a knife or cake tester comes out clean.
8. Cool the custard cups on a rack for about 20 minutes, then refrigerate. Serve when the custard has completely chilled.

# Mini Pavlovas with Whole Egg Lemon Curd

*Yields 12 servings*

Pavlovas are meringue crusts that can be filled with whipped cream, fruit curd, fresh fruits, ice cream, edible flowers, and pretty much anything else your imagination whips up. You can make one or two big shells, or smaller individual shells, like we did for our garden party. Pavlovas make a light and very pretty dessert, especially for spring and summer.

If you want to shape your pavlova shells instead of going free-form, you can draw your intended shape on parchment paper and pipe the meringue into the outline. You can also use piping tips for an even fancier finish! Otherwise, spoon meringue into roughly the shape and size you want. Then use a spoon to make an indentation in the center of each shell, basically forming a shallow pie shell.

Our lemon curd may be the easiest you'll ever make. No need to separate your eggs, no double boiler, and if you get cooked egg bits, you can put it through a sieve or strainer and it will end up perfect. It's possible that we've been known to stand around and eat this curd with a spoon.

*Free-form pavlovas garnished with raspberries.*

### *Meringue Shell Ingredients*

- 4 egg whites
- 1 cup granulated sugar
- 1 teaspoon vanilla extract
- 1 teaspoon cornstarch
- 1 teaspoon lemon juice

### *Meringue Shell Instructions*

1. Preheat oven to 200° F.
2. Line one or two baking sheets with parchment paper or silicone mats.
3. Beat the egg whites on high speed. As they begin to form soft peaks, begin gradually adding the granulated sugar, a tablespoon or two at a time. Continue beating and adding sugar until the whole cup has been incorporated and the egg whites are shiny and hold stiff peaks.
4. To form your meringue shells, pipe or spoon the beaten egg whites onto the prepared baking sheet(s). For mini pavlovas, you'll want to make them around 3–3.3 inches in diameter.
5. Bake for 1.5–2 hours, checking for doneness every 10 minutes. When finished baking, the shells will be slightly browned and slightly cracked.
6. Turn off the oven and leave the shells in the oven overnight.

### *Lemon Curd Ingredients*

- ½ cup fresh lemon juice (we use freshly squeezed, but the curd is still delicious with bottled juice)
- ½ cup sugar
- 3 large fresh eggs
- 6 tablespoons unsalted (or dairy-free) butter, cut into chunks

### *Lemon Curd Instructions*

1. Whisk together juice, sugar, and eggs in a medium-sized saucepan.
2. Add butter and cook over moderately low heat, stirring almost constantly, until the curd is thick enough to hold marks of a whisk or spoon across the bottom of the pan, about 5 or 6 minutes.
3. Transfer lemon curd to a bowl and chill before using.
4. Fill the cooled meringue shells with lemon curd and whipped cream.
5. Garnish with fresh fruit—raspberries are perfect—or edible flowers to make the pavlovas extra pretty. Enjoy!

*Salad in a vintage Indiana Glass Company bowl.*

# Shrimp and Egg Cobb Salad

*Yields 6–8 servings*

A fast, easy, and delicious main dish salad that is perfect for warm weather and uses up a lot of your spring and summer bounty of eggs.

## Ingredients

About 1 pound of fresh spring greens, washed and dried

1 cucumber, peeled and sliced in half rounds

8 ounces cherry tomatoes, sliced in half

1 can of corn, rinsed, or 1 cup fresh or frozen corn

1 can of black beans, rinsed

1 avocado, cut into chunks

1 jar of sweet corn salsa

Vinaigrette dressing

½ dozen hard-boiled eggs (or more, if you prefer)

½ pound of steamed or cooked shrimp, peeled, tails removed, seasoned to taste

## Instructions

1. In a very large, wide salad bowl, add all of the greens. Layer cucumber and tomatoes on top, followed by corn, beans, and avocado.
2. Reserve about ⅓ of the sweet corn salsa, then layer the rest over the bowl.
3. Drizzle the vinaigrette over it.
4. Add the shrimp and eggs on the top of the salad, salt and pepper to taste, and then add the last of the sweet corn salsa to garnish.

*The darker the yolks, the more golden the cake.*

# Golden Yolk Cake

*Yields 12 servings*

Golden Yolk Cake is an easy and delicious cake baked in a tube or Bundt pan. You will use 12 chicken egg yolks (see recipe for conversion to duck or goose eggs), and you'll have the whites left over for meringues or egg-white omelets. (The egg whites also freeze well for later use.) This cake has a spongy texture, somewhere between a pound cake and an angel food cake. We love it served with fresh strawberries and whipped cream, but you can also split and fill this cake with fruit curd, or decorate it with glaze and edible flowers. As always, we include gluten- and dairy-free substitutions.

## Ingredients

- 3 cups all-purpose flour (or gluten-free 1-1 flour)
- 2½ teaspoons baking powder
- ½ teaspoon salt
- 12 large egg yolks at room temperature (or 10 duck yolks or 4 goose yolks)
- 2 cups granulated sugar
- 1 cup milk (or plant milk), cold
- 1 teaspoon pure vanilla extract

## Instructions

1. Preheat oven to 350°F.
2. Spray or grease a tube pan or Bundt pan with nonstick spray.
3. Whisk flour, baking powder, and salt together in a medium-sized mixing bowl.
4. In a large mixing bowl, beat egg yolks until very fluffy and thick.
5. Using medium/high speed, gradually beat in the sugar, stopping to scrape the bowl as needed.
6. Reduce speed to low and add the milk and vanilla extract. Beat until incorporated.
7. In two separate additions, add the flour mixture while beating on low and scraping the bowl when necessary. Beat only long enough each time to blend, about 2 minutes.
8. Pour batter into the prepared pan.
9. Bake 50–60 minutes, until lightly golden, or until a knife or cake tester comes out clean.
10. Cool cake for 15 minutes on the wire rack before removing from the pan. Turn the cake out and cool completely.

*We used a vintage goose pâté dish to serve this easy and delicious appetizer.*

# Shrimp and Egg Pâté

*Yields 20 appetizer servings*

We made this Shrimp and Egg Pâté appetizer with New Year's Eve in mind, but it's great for any time! This is essentially a chunky dip, but it's thick and works best spread on crackers or veggie slices. A better description may be something between pâté and terrine. We packed it into a mold (a small GladWare container) and served it on a vintage goose appetizer plate. But you can also roll it into a log, or even just dump it onto a plate and surround it with slices of baguette. You can customize the spices, and use a different veggie in place of the roasted red peppers if you'd like.

## Ingredients

2–3 hard-boiled eggs, peeled

1 cup cooked shrimp, tailed and deveined

½ cup roasted red pepper

4 tablespoons (2 ounces) butter (or dairy-free butter), cold

4 tablespoons (2 ounces) cream cheese (or dairy-free cream cheese)

1–2 teaspoons Old Bay seasoning

½ teaspoon dried dill

## Instructions

1. Combine all ingredients in the bowl of a food processor.
2. Pulse until the mixture looks the way you want it. We kept it a bit chunky, more like a terrine. But you can process it down into a smooth pâté if that works better for you. You can also chop with a knife and stir everything very well together in a mixing bowl if you prefer.
3. Spray or oil whatever container you want to use as a mold. You can also line your dish or mold with plastic wrap that has been sprayed.
4. Pack in the pâté, smooth the bottom, and place in the refrigerator to chill for at least an hour.
5. Unmold onto a serving plate or platter. Serve with crackers or slices of baguette and small knives for spreading. Voilà!

*The first recipe that we featured on the podcast!*

# Potato Noodle Stew with Eggs

*Yields 4–6 servings*

This delicious stew is loosely based on a traditional Spanish dish. It is hearty and delicious and cooked entirely in one pot; it tastes even better the day after it's made. Whole eggs, in the shell, are lowered into the pot to cook as the stew simmers. You can omit that step and simply chop up hard-boiled eggs and serve them on top of the stew if that's easier, or if you are feeding someone with a compromised immune system.

## Ingredients

- 2 cubanelle peppers, seeded and diced
- 1 onion, diced
- 3 garlic cloves, chopped
- 1 28-ounce can of diced tomatoes, *un*drained
- 3–4 medium potatoes, peeled and cubed (Yukon Gold work well)
- 3 cups of broth (we use veggie, but any kind works)
- 1 handful of parsley, coarsely chopped
- ¾ teaspoon ground cumin
- 1 pinch of saffron threads—about ⅓ of a teaspoon
- 2 teaspoons salt (we use sea salt)
- ¼ teaspoon black pepper
- ⅓ box of spaghetti (or gluten-free spaghetti), broken into 2-inch lengths
- 6 raw eggs still in the shell (washed well!)

## Instructions

1. Heat a generous splash of olive oil in a Dutch oven or large pot over medium heat.
2. Add peppers, onion, and garlic and cook, stirring often, until soft and starting to brown.
3. Add the undrained can of diced tomatoes and allow to simmer for about 10 minutes, stirring occasionally.
4. Add potatoes, broth, parsley, and all seasonings.
5. Bring the stew to a boil, turn heat to low, and simmer, covered, until the potatoes are fork-tender, about 20–25 minutes.
6. Add pasta, stir, and allow to simmer over low/medium heat for 10 minutes.
7. Using a ladle, lower the eggs into the stew and simmer for 15 minutes to hard-boil them.
8. Remove the eggs, peel, chop, and sprinkle over top of bowls of stew. Or dish out the soup with the eggs still in and let everyone peel their own. Enjoy!

*Eggs are a great source of protein in stir-fry.*

# Thai-Inspired Pumpkin and Egg Stir-Fry

*Yields 2 servings*

## Ingredients

1 small pumpkin or kabocha squash, about 2.5 pounds

3 large garlic cloves, minced

1 tablespoon fish sauce

1 tablespoon soy sauce tamari (or gluten-free soy sauce)

1–2 cups of water

4 eggs, lightly beaten

¼ teaspoon ground pepper, white or black

Olive oil

## Instructions

1. Peel the squash and cut into small cubes, around 1-inch pieces.
2. Heat the oil in a large skillet over medium heat. Add the garlic and sauté for a minute or so.
3. Add in the squash, fish sauce, and soy sauce, along with enough of the water to cover the squash. Bring to a boil and allow the squash to steam until it is cooked through and the water has almost evaporated. This will take about 12–15 minutes.
4. Push the squash to the sides of the pan. Pour the eggs into the center and stir them rapidly until they're just cooked, about 2 minutes or so.
5. Stir everything together and season with salt and pepper to taste. Serve over steamed rice or rice noodles and enjoy.

*Upgrade your deviled eggs with some of Maryland's traditional ingredients.*

# Maryland-Style Deviled Eggs

*Yields 12 servings*

## Ingredients

6 hard-boiled eggs, peeled

¼ cup mayonnaise

1 teaspoon Dijon mustard

1 teaspoon–1 tablespoon Old Bay seasoning, according to taste

Salt and pepper to taste

A dash of hot sauce *(optional)*

½ cup lump crabmeat

## Instructions

1. Cut each egg in half and scoop out the yolks.
2. In a small or medium mixing bowl, mash the yolks together with the mayonnaise, Dijon mustard, Old Bay, salt and pepper, and optional hot sauce. Mix well.
3. Spoon or pipe filling into the egg halves.
4. Arrange eggs on a platter or egg plate. Top each egg half with a pinch of crabmeat. Sprinkle a bit of Old Bay on top and serve.

# Vegetable Soup with Egg Dumplings

*Yields 4–6 servings*

This recipe is loosely based on a Hungarian dish, Paprika Spiced Cauliflower Soup. We don't use Hungarian paprika or hot Hungarian paprika, just regular paprika, and not as much of it. But we do use carrots, cauliflower, and egg dumplings. You can make the dumpling dough ahead of time and pop it in the fridge while you make the soup. You will want a rich, tasty broth since this is a fairly simple soup. You can also add other veggies and peas or beans for some protein.

*Egg dumplings make this vegetable soup something special.*

## Ingredients

*For the dumplings*

- ⅓ cup all-purpose flour (or gluten-free 1-1 flour)
- ½ teaspoon of salt (or just a pinch, if using salted butter)
- 2 tablespoons butter (or dairy-free butter), cut into small pieces
- 1 large egg

*For the soup*

- 4 tablespoons butter (or dairy-free butter), or 2 tablespoons butter and 2 tablespoons olive oil
- 1 teaspoon of paprika
- 1 medium onion, finely chopped
- 6 cups broth or rich stock of your choice
- 1 small head of cauliflower, cut into small florets, or a 12-ounce bag of frozen florets
- 2 medium carrots, peeled and sliced

## Instructions

*To make the dumplings*

1. In a small mixing bowl, whisk together the flour and salt.
2. Using a pastry blender (or your fingers), cut the butter into the flour until it looks like pieces of coarse meal.
3. Add the egg and stir until a sticky dough forms.
4. Cover the bowl and pop into the fridge while you make the soup.

*To make the soup:*

1. Place a large soup pot over medium-high heat and melt the butter / olive oil.
2. Add the onion and paprika and cook, stirring frequently, for around 5–6 minutes, or until the onion begins to soften.
3. Add the broth and veggies and bring to a boil. Reduce the heat to medium/low and cook until vegetables are tender, about 15 minutes. Make sure you stir every so often so that nothing sticks to the bottom of the pot.
4. Use a regular teaspoon to scoop and drop the dumplings into the soup. It's sticky dough, so you may need to push them off and into the pot with a second spoon. The dumplings are not huge, and they float on the top of the soup and cook fairly quickly. It usually takes about 3–5 minutes for each one to cook.
5. When the soup is ready, serve and enjoy!

*See ya 'round the coop!*

# RESOURCES

**Coffee with the Chicken Ladies**
https://coffeewiththechickenladies.com/

### *Hatcheries*

Greenfire Farms
https://greenfirefarms.com/

Murray McMurray Hatchery
https://www.mcmurrayhatchery.com/

My Pet Chicken
https://www.mypetchicken.com

### *Feed*

Nutrena NatureWise Feeds
https://nutrenaworld.com/products/poultry/

- All Flock
- Feather Fixer
- Harvest Blend (whole-grain)
- Hearty Hen
- Layer Pellets/Crumble
- Silkie and Ornamental Layer Feed

### *Treats and Other Useful Products*

Crazy K Farm (Crop Bras and Diapers)
https://www.crazykfarm.com/

Grubbly Farms
https://grubblyfarms.com/

Lisa Steele's Fresh Eggs Daily Products
https://fresheggsdailyshop.com/

Strong Animals Chicken Essentials
https://www.getstronganimals.com/

Tractor Supply Company
https://www.tractorsupply.com/

### *Plastic Coops and Runs*

Nestera
https://nestera.us/

Omlet
https://www.omlet.us/shop/chicken_keeping/

### *Helpful Websites*

American Poultry Association
https://www.amerpoultryassn.com/

Association of Avian Veterinarians
https://www.aav.org/

The Livestock Conservancy
https://livestockconservancy.org/

Poultry DVM
https://poultrydvm.com/

# INDEX

# ABOUT THE AUTHORS

**Holly Callahan-Kasmala** and **Chrisie DiCarlo** are the creators and cohosts of the *Coffee with the Chicken Ladies* podcast. Best friends for more than forty years, they have worked side by side throughout their lives and started the podcast to help chickens and their people navigate their lives together. By applying their own decades of experience and helping their listeners through all kinds of chicken care situations, they distill the basic, practical information that beginning chicken keepers need to succeed.

Holly grew up on a horse farm and has a lifetime of experience caring for livestock. She has kept chickens for more than twenty-two years and shares her wealth of practical experience with chicken keepers of all levels. She also holds an MA in historical studies and specializes in livestock history. Holly worked as Assistant Curator of Manuscripts at the Johns Hopkins University Archives and as the Local History and Genealogy Librarian at the Maryland Center for History and Culture. Her love of history, poultry, and meticulous research combine to add depth and historical context to all aspects of chicken keeping. She lives in Millers, Maryland.

Chrisie spent fifteen years as a vet tech/trauma nurse and client educator in a busy urban Baltimore City animal hospital. Her experience working with both clients and staff makes her uniquely qualified to help chicken keepers understand anatomy and care of backyard chickens. Chrisie guided her daughters through raising their first chickens together, giving her valuable insight into the benefits of chicken chores on children's development. In addition, twelve years with her own flock of chickens brings added layers of experience to her writing. She lives in Fallston, Maryland.